GETTING MORE FOR LESS

The Gravity of America's Choices

GEORGE LAROQUE III

AuthorHouse™
1663 Liberty Drive
Bloomington, IN 47403
www.authorhouse.com
Phone: 1-800-839-8640

Published by AuthorHouse 10/06/2014

ISBN: 978-1-4969-4497-9 (sc)
ISBN: 978-1-4969-4496-2 (e)

Library of Congress Control Number: 2014917985

This book is printed on acid-free paper.

To every thing *there is a season, and a time* to every purpose under the heaven.

—Ecclesiastes 3:1 (King James Version)

To every one *there is a reason and a rhyme* for every choice here on earth.

Contents

Tinker Tailor

Daisy, Daisy, who shall it be?
Who shall it be—who will marry me?
  -Tinker, Tailor,
  -Soldier, Sailor.
  -Rich man, poor man, beggar man, thief,
  -Doctor, Lawyer, Merchant, Chief.
Daisy, Daisy, what will I be?
  -Lady, baby, gypsy, queen.
Grandmother, Grandmother, what shall I wear?
  -Silk, satin, cotton, rags.
How shall I get it?
  -Given, borrowed, bought, stolen.
How shall I get to church?
  -Coach, carriage, wheelbarrow, cart.
Where shall we live?
  -Big house, little house, pigsty, barn.
How many children shall we have?
  -One, two, three, four ...

—American childhood nursery rhyme about the choices we make and the power of free will...

Introduction

A Reason for Our Lives: Choices and Consequences

"One, Two, Three, Four ..."

I don't believe anything happens by random chance. There are no coincidences. Everything is a consequence of some other action in our universe. I believe everything follows a master plan, and therefore everything happens for a *reason*. This is true for both heaven and earth. Events in our earthy lives could follow God's plan or more simply our own choices, as implied by the childhood rhyme "Tinker Tailor."

I believe everything (and hence everyone) obeys predictable laws of the universe,[1] including not only the simpler predictable occurrence of a *season* but also the meaningful use of our *time*. And while we may not always understand the *rhyme* or the *reason* behind everything and for everyone, I believe in quantitating laws and uncovering rules whenever possible within my small frame of reference. We all have purpose and value, and it is not a random process for any of us.

I am blessed to be a father of four children, all of whom are modern-day equivalents of Daisy from the nursery rhyme. They each currently face similar vitally important decisions that they mostly have to choose for themselves. These choices include the extent of their education, their future careers, and their (hopefully distant) choice of a spouse. They will

---

1 This is from Aristotle's philosophy that emphasizes a purpose to everything.

have to decide where they will ultimately live, how they will perceive and help others, and even how many of their own children to one day have within their own families, thereby continuing life's repeating cycle. I am able to help shape their decisions and the resulting outcomes—both good and bad. I now realize that these decisions and outcomes obey laws of nature—*more or less.*

My children's decisions often result in outcomes that may seem random, like the simple throw of a dice, or that seem unfair to me as a parent, like how my four children have chosen to adopt different core values within our family of origin. For example, how and why one of my children has chosen a life of chemical addiction seems unimaginably random and unfair. But nonetheless there are logical reasons for this skewed behavior.

There are in fact hidden rules and principles that predictably shape our thoughts and disproportionately affect our actions. These rules influence everything we assign value to in the game we perceive as life. By uncovering for my children a few of these vital rules, coupled with a few self-empowering skills, I hope to help them navigate through the fractal maze of difficult life choices. As Joseph Ford said, "God plays dice with the Universe. But they are loaded dice. And the main objective is to find out by what rules they were loaded and how we can use them for our own end."[2] I am convinced that the dice *are* mysteriously rigged, yet I also believe the preloaded rules can be learned early in life if you pay attention.[3]

I wish the secret were as simple as telling our children to (1) adopt meaningful core values, (2) get a quality education to the best of their individual ability, (3) excel in whatever they want to do, (4) pick something they passionately enjoy as an occupation, (5) attract others of similar values throughout life, (6) marry someone with similar core values, and (7) pick a place to live that is conducive to well-being.

---

2 Richard Koch, preface to *The 80/20 Principle: The Secret to Achieving More With Less* (New York: Doubleday Publishing, 2008). Joseph Ford's quote from the preface relates to an Einstein quotation that says, "God does not play dice with the universe"—suggesting that everything follows predefined rules and nothing is left to chance in the universe.

3 I would point out the most efficient way to learn how they are loaded is to get to know God.

But alas, it is not so simple, as you may already realize. First of all, that is too much to remember for anyone. Secondly, times change, and perhaps what I now think is important based on my early life experience differs within the modern world and relative to the newer generation. Thirdly, *telling* people to do something rarely ever works, and as a parent I have found this to be especially true with my kids. Modeling it is more fruitful—as an individual, a family, a community, and as a nation—and perhaps this is where I (and we) have been most deficient.

Instead of overwhelming my children's young developing minds with many trivial life lessons, therefore, I have made the choice to teach and model a few valuable skills that can help my kids think more analytically about what they can do to succeed in life while using less effort. I want to simplify it all for them. *I want to teach our kids to use less effort to achieve the things they most desire*. Yes, you read that right! And this can be quite achievable if they can learn several simple ideas that they already know and if they can change the way they have been conditioned to think critically about our popular culture.

While adopting one of my foreign-born children, I learned the proverb "To give your child a skill is more valuable than a thousand pieces of gold."[4] Since my wife and I have been blessed with *four* children, I thought it would be economically more efficient to give them each a few valuable skills, knowing they will be richly rewarded, albeit not with gold. While I have not successfully achieved this outcome equally with all four of my children, most of them are able to get it, and to borrow a phrase, "I am still learning"[5] as well!

Within my own life experience, there has indeed been a learning curve with each child—*"one, two, three, four."* My wife and I made a few early mistakes with our first child, and we have hopefully learned from these mistakes through proper feedback and change. With the second child we

---

4 This is an anonymous Chinese proverb.

5 This is a quote often attributed to Michelangelo (whether erroneously or not), who accomplished much by the time he was in his twenties. I am still learning with my own children, and we all need to emphasize continual learning. *Wikipedia*, s.v. "Michelangelo," last modified July 28, 2014, http://en.wikiquote.org/wiki/Michelangelo.

learned a bit more and achieved a bit more. By the third we were getting better, and by child number four we are about as efficient as we can get as parents, although we are still far from perfect. Our children's life stories in fact have been imbalanced, similar to most outcomes in the human experience we each call life. That imbalance in their childhood outcomes was a huge factor in what humbly motivated me to write this book. I was looking for answers not only for them but also for myself as their parent.

For most Americans, including my four children, life *does* seem to haphazardly fall into place at times, much like a game[6] of random chance, with many diverse outcomes that at times seem both inequitable and unfair. Who your parents are, where you live, and during what time period you are born are just a few examples of some of the random life events that seem beyond your control—and yet these variables can profoundly and disproportionately shape many of your future outcomes. I believe awareness of a few predictably imbalanced but interconnected rules that I share here can and will help you learn to reshape your future.

I have learned a few valuable lessons in my life that I feel are worth sharing with my children. These lessons represent tiny but powerful buried principles that I have unearthed through my own personal and sometimes dirty and less-than-perfect life experiences. A few secret rules invisibly direct this inequitable game we call life.[7] Why no one formally teaches us these concepts as young adults is unclear to me. Perhaps if we did so, we would in theory realize how *inefficient* we truly are as parents or as

---

6 A branch of sociology known as *game theory* evolved from similar concepts of treating human behavior as games (like checkers and chess) and mathematical probabilities. By no coincidence the rhyme that I chose for the preface has origins from a monk, who linked human morals and ethics to the various careers (like tinkers and tailors) and their ranks (and the implied power) through the different game pieces used in the game of chess. His moral treatise was known as *The Game and Playe of the Chesse* and was one of the first two books translated into the English language in the 1470s (from Italian Jacopo Da Cessole, translated by William Caxton).

7 All my children have seen and played the Milton Bradley game of Life, more recently in apps on their electronic tablets. The game was created in the 1860s when the same core values were emphasized in our education, marriage, and family. The original version required skills of counting and reading just like the modern version. However, the modern versions add an element of chance and include rewards for taking risk and for helping others (e.g., the less fortunate, such as the homeless). Life remains one of the more popular board games from American history.

educators. Understanding the nature of these peculiar imbalances is crucial if we want to survive in this skewed world and even more so if we hope to collectively make changes that can both disproportionately favor our families and help improve American society.

As a *doctor* who has earned high levels of both income and education, I have observed a certain peculiar ubiquity of imbalanced "popular" outcomes throughout my own life experience, including the obvious inequality of American income distribution, the skewed choice of American occupations, and the lopsided amount of educational achievement. Furthermore I have observed how each one of these can disproportionately affect not only America's wealth but also our country's health and general happiness. What strikes me as more relevant is the surprising fact that all of these "American" imbalances are neither unique to my lifetime experience nor to my all-American geography. These results are discoverable everywhere. They define the human condition, and they can be observed to follow both a rhyme and a reason.

Humans categorically structure through rank everything within our lives through seemingly random order (numerically or otherwise), as we try to add some semblance of reason to the choices throughout our everyday lives. While none of us is any better than another, we are constantly ranking ourselves in comparison to others through our various gifts and abilities, in our occupations, in our lifestyles, through our choice of religion, our popularity in the clothes we wear, our athletic and academic achievements, and so on. There is actually a rhyme and reason to the seemingly random rank, and that is in part what I hope to illustrate here.

I want to enlighten my children by revealing to them that what is often popular (i.e., in this way we rank outcomes) in American culture is not always what is best for them, and certainly it is not random. *More-popular* outcomes are often *less desirable*, especially if you add in the measures of time, quality, and work (effort) as can be seen through the benefit of hindsight. Understanding that this reciprocal and very nonlinear relationship creates natural but recurring *imbalances* is paramount in your early education if you wish to magnify these effects with the power of your limited earthly time. Learning what to purposefully rank and value within your lifetime therefore can be even more disproportionately rewarding if you can learn a few concepts early.

A few *principles of imbalance* that I have discovered are not exclusively American patents. Although invisible at first, they can be found everywhere, especially anytime human choice is involved. They are often the result of predictable interactions between our natural abilities and the opportunities offered by nurturing these abilities from our social and physical environment (i.e., nature and nurture). While we are currently incapable of modifying our genetic constitution in the short term, we are able to change the constitution[8] of our culture. Therefore what I intend to show my children in their own lives and what I hope to model to them as a parent is that *we are all capable of change through the power of our choices.* This book emphasizes how we *choose* to interact as social beings within society (America is my frame of reference) using a powerful tool that we have all been gifted—*free will.*

Every society, no matter where or when it develops, eventually evolves into its own self-organized system (e.g., America) that exists in some form of a dynamic equilibrium. With time, each society evolves its own particular clustered ideas and attitudes based on that unique cultural mind-set. "A social mind is the cluster of ideas and attitudes that gives to a society whatever uniqueness or individuality it may have as an epoch in the history of thought."[9] A country (or state or city) essentially operates and changes as a dynamically evolving organic machine with a certain characteristic efficiency—with some countries (or states or cities) becoming more efficient in certain measures (e.g., equal rights and more opportunities, as in a "free" country such as America) and less efficient (e.g., poor health) in others. As soon as we form a group of collectively more than one (the family is the beginning of a fractal society), we become naturally imbalanced through this process of efficiency

---

[8] We do pride ourselves on our ability to spell out our desires culturally in the US Constitution. I don't mean to imply we should rewrite it, but I do mean to suggest we should rethink what is valuable to each of us within the same context. I also imply we may be able to change our genetics at some point in the future.

[9] Stow Persons, preface to *American Minds: A History of Ideas* (Holt, Rinehart and Winston, 1958).

(or lack thereof), and learning just how much and what we can do to reshape this imbalance within our skewed human experiences defines much of the challenge between the start and finish to this game of life.

Human beings are fundamentally similar everywhere and always, and therefore I believe the subtle differences between our many outcomes are *not* mostly genetically determined[10] but are disproportionately the product of our environment and the prevailing cultural influences.[11] These internal efficiencies are in fact guided by the nature of each culture (our nurture) and how society selectively rewards or penalizes these inputs (our nature) through a process of negative and positive feedback (like income and wealth) with each of us. This feedback forms the basis for what we then consider as values and beliefs within our nation—such as with American popular culture or within smaller groupings of people like the individual household family unit.

Getting more or *less* is really all about the essence of our life experiences since nothing we do is ever absolute. Most everything in life and within our universe is experientially relative, and most of us never quite figure out what that frame of reference is comparable to or what we are here *for*.[12] Maybe you faithfully believe in an absolute reference with your own choice in *God* through your religion; or philosophically you believe it is unconditional *love* and universal tolerance of our natural differences

---

[10] I say this to silence the extremists who would use the idea of genetic superiority to explain most of our differences. Although genetics clearly play a role in certain abilities, we are mostly all the same skewed human creatures genetically endowed with a similar DNA alphabet base. The rest is shaped and reshaped by us and through culture and how we dynamically interact within that culture to communicate using our genetically similar letters to form unique and common words.

[11] If this seems remotely Darwinian, it is not a coincidence. Darwin similarly noted an unusual distribution of people and animal populations that seemed to mirror their environment, suggesting to him a modifiable and adaptable aspect to all life-forms that led to his law of natural selection. In essence we humans are all able to change our behavior and adapt in similar ways, which is the premise of this book. If you believe you have the power to adapt and change to your local environment, then you too are thinking in Darwinian terms.

[12] Finding your frame of reference (FOR) is part of the main message in this book. It is what we all seek throughout our lives, relative to others in our human relationships or relative to a higher power through religion. Either way, these rules apply, and I hope to help teach you how to more or less find your own frame of reference in your life. That is why you are here. But it is not as easy as it sounds. We are all in a way lost, relatively speaking, and all searching to find a way back home to our original frame of reference.

and (human) abilities; or scientifically you seek absolute *truth* through knowledge of a grand unifying theory that explains everything within the vast universe. Regardless of our frame of reference, we each more or less struggle with similar fundamental concepts that define for each of us the imbalanced human outcomes we individually experience. As Vilfredo Pareto, an early social engineer, eloquently summarized life, "It is always a question of a more or a less."[13]

For most of us, money is the visible American model that mirrors this perpetual imbalance, reflecting the nature of the complex human struggle between a *more* or a *less.* Many of us can relate to the obvious imbalanced findings of income distribution within society—fewer people getting more, a.k.a. the *rich man*, and more people getting less, a.k.a. the *poor man*. For this reason, I will use income distribution as a common measure throughout this book to demonstrate some of these relationships. Through the readily visible examples of income inequality in America, I hope to generalize how we can make appropriate change to any human input[14] in order to reshape the outcomes more (or less) favorably.

This book, however, is not meant to be mostly about income disparity within America, although money can be an efficient tool to teach us a few basic principles. The message more generally is one about the universal mechanics of other human behavioral imbalances, including our differing choices in education, our skewed choices in careers, and the hidden patterns in our marriages and similar values. I plan to uncover the general forces that profoundly shape family values and the influential factors linked to our geography, our religion, and other cultural ideas to which we all gravitate by attaching rank and value.

Numbers are my own internally programmed way of communicating these imbalances,[15] and therefore it seems logical that I use counting ("*one,*

---

[13] Pareto, *The Mind and Society*, ed. Arthur Livingston (San Diego: Harcourt, Brace and Company, 1935), 1:57.

[14] Money distribution and people distributions are just different collective groupings that follow similar principles. People follow money, and money follows people. The ideas are learned and propagated with time (and a rhyme).

[15] I say this since my nickname for all my life has been "Trey," which means three on a die or in a game of poker (three of a kind).

two, three, four") and money through the simple power of a nursery rhyme as my preferred tool, especially in light of my own *four* children.

Money is a useful and powerful tool. It is efficiently used to get something we desire in relation to others, and it is a tool we all seem to understand in relative terms. Much like the disproportionate distribution of money within America, our general behavioral outcomes with respect to each other are really no different. We are just as skewed in our values and our behavioral choices as Americans, both of which are no product of random coincidence. I plan to show you in this book (1) just how skewed we are through a few repeating principles, (2) what this means for you as an individual, (3) how you can use this knowledge to your advantage, and (4) how we can collectively learn from this as a nation and favorably shape outcomes by understanding the powerful link between America's choices and their consequences.

I am no lawyer, but I have learned a few vital laws that summarily describe human behavior and demonstrate a few cyclically repeating imbalances that are fundamental life theorems worth knowing. They can be useful in all areas of our lives, including our health and wealth. One of these "laws" is by no coincidence aptly referred to as *the law of the vital few.*[16] Most people don't know of it, and even fewer understand fully how to use it. Secretly, a few similar *natural laws* help shape important societal (moral, economical, philosophical, and religious) outcomes. These laws are teachable, and my children can use them to achieve the vital results they each desire (with money or careers, in marriage or education, and with religion or family values). More broadly as a society, Americans can use these same relevant laws as tools to help reshape the growing disparity within our greater population that is being created by the very nature of *popular* societal choices and outcomes.

In the end, I hope to show my own figurative Daisys how to anticipate the natural consequences of their current life choices now and well into

[16] Joseph Juran, another engineer, introduced this term, and his law is the subject of part 3 of this book. Ironically he also became a lawyer. In summary, his law says that a few vital causes disproportionately account for most of the results we are interested in achieving. My translation is "life is skewed" (or for my younger kids, "life is unfair").

their future. I wish to impart to them the relevance of adopting core values that are consistent with healthy, productive lives at home within our family of reference, their own future families, and American society. Most importantly, I want them to better understand the available tools to help them efficiently navigate the many choices they now face in their nonlinear and nonrandom world by learning to think critically and creatively.

We are all here for a reason, and there is no better time or season for us to learn these valuable principles. The reason coupled with a simple childhood rhyme can teach each of us—child and adult alike—how to better navigate the game of life, use its (or God's) rules more efficiently, and hopefully improve American society along the path of our unique personal journeys.[17]

[17] In game theory, this would be known as a *win-win* outcome, and this is a skill worth learning early in your life. Make your life full of win-win outcomes, and everything else will naturally follow.

"Daisy, Daisy, Who Shall It Be? Who Shall It Be—Who Will Marry Me?"

Learning to Ask the Right Questions

This book has a dual purpose—*a rhyme* and *a reason*. It is about more than just selfishly teaching my kids about a few vital choices in their lives, such as what they should do, where they should live, and whom they should marry—although these tips certainly can be useful. It is more than just citing the many ways America is becoming severely imbalanced because of what is becoming more popular in our children's culture. These concepts can be daunting to our children. This book more relevantly seeks to teach them a repeating association between not just these simpler choices but also *all* of the choices we make in our everyday lives and the related predictable life outcomes that seem consistently imbalanced for everyone. This book teaches them to think critically[18] and to ask the relevant questions while understanding just how skewed the results can become.

To satisfy the first purpose, I plan to review a few helpful tips that I have learned over my short lifetime. These useful life skills include how to think critically and examine one's own life choices. I integrate a few examples from my family, linked through the invisible power communicated by a popular children's nursery rhyme from my own childhood experience.

To address the second goal, I want to demonstrate how anyone can use what are known as *power laws* to get favorable outcomes. A little (*less*) knowledge can be used to get *more* quality outcomes throughout our lives—whether that is getting more money, more happiness, higher education, a

[18] The three Cs are critical thinking, creativity, and continual learning. Arthur Levine and Diane R. Dean, *Generation on a Tightrope: A Portrait of Today's College Student* (San Francisco: Jossey-Bass, 2012).

greater connection with God, better health, more-equitable laws (e.g., tax laws), a better career, or simply more leisure and quality time. Power laws are mathematically skewed tools that help explain the imbalances we all experience in our lives. They are subtle, hidden principles that secretly govern much of the mechanics of what happens throughout our lives, and I plan to reveal them to you.

Learning these principles and more can be achieved using *less effort and time* once you understand the reasons for these outcomes, which can be taught by a simple children's counting nursery rhyme I use from the start. This tool of efficiency is the basic essence of *Getting More for Less.*

I purposefully chose each and every word in this book, including the nursery rhyme, for a reason, as we all unconsciously do in order to communicate with others. The classic American nursery rhyme that I learned to repeat as a kid, commonly known as "Tinker Tailor," can be helpful as a simple communication tool, and I will use it to demonstrate most of the repeating imbalances I have observed within my lifetime—not just in content but in style and mechanics as well. On a magnified scale, these same few reasons for how, why, and what we choose to do (through our efforts in communication or otherwise) are applicable[19] to any collective society and can be used to culturally reshape larger population outcomes, such as social and economic inequalities, which can greatly affect general well-being and happiness through simple links between choice and consequence.

My own life experience has taught me that most outcomes of any significant value involve some aspect of human choice (whether it is our choice or someone else's) through exercising our free will. Rarely do we get anything we want without doing something to get it. While not everything or every reason is intuitively obvious to everyone, we can usually modify our choices by learning something from the naturally linked consequences.[20]

---

[19] These laws possess a special property known as scale invariance, which means they apply to small numbers or large populations. In fact, the relationships often become more obvious the larger the numbers.

[20] I like Einstein's definition of free will: "I do not at all believe in free will in the philosophical sense. Everybody acts not only under external compulsion, but also in accordance with inner necessity. Schopenhauer's saying, 'A man can do as he wills, but not will as he wills,' has been a real inspiration to me since my youth; it has been a continual consolation in the face of life's hardships, my own and

Often this choice-consequence relationship is immediately obvious to our children, like with the simple example of touching a burning-hot stove. But at other times this natural relationship is subtler, like the complex issue of income disparity within America, and it takes decades for our brains to process and learn these outcomes, which partly explains the observed imbalances within our society.

There is an inherent dependence on the power of time and space with many of these relationships,[21] and for this reason, I do not expect my children to immediately understand the nuances of these skewed interconnections when they are young. However, as they mature and experience these repeated cyclical interactions of choice and consequence more generally, I hope they will learn to appreciate the nature of these peculiar reciprocal relationships in such a way that they can enjoy efficient outcomes that mostly favor their continued well-being.

I believe the universe is naturally skewed but in an orderly way that favors efficiency. I have been patterned to think of life as regularly irregular.[22] By a similar earthly view, I maintain that mankind is naturally skewed in our programmed[23] ability to achieve whatever endpoint we may desire. While many Americans prefer to believe that our nation was founded on the ideals of social equality (of rights and opportunities, for example[24]), what many of

---

others', and an unfailing wellspring of tolerance." Walter Isaacson, *Einstein: His Life and Universe* (New York: Simon and Schuster, 2007), 391.

21 Some of these skewed relationships mathematically have a time component. For example, *compound interest* is a valuable relationship that allows you to magnify consistently good habits disproportionately (very nonlinearly) over time, as in investments and savings. This is why people with good habits early become *more* successful sooner, for longer periods, and with seemingly *less* effort. *Space and time* represent all the dimensions with which we are familiar in our common thinking.

22 This is a nod to my medical background where a pattern of irregularity repeated at regular intervals can be diagnostic of certain heart arrhythmias. Patterns are prevalent everywhere, and looking for them is part of this process I allude to with critical thinking.

23 By this I infer our DNA is incredibly skewed within the elements of our planet, which are incredibly skewed in relation to the elements of the universe, which are unfathomably skewed in relation to everything else. It is a mind-boggling skewed skew of skews from the big bang on, which has all favored our existence here and now.

24 These are the potential inputs that we can make more equally available, although the outcomes will still naturally become skewed based on our different abilities or, as I prefer, *efficiencies*. I would point out that slavery and women's rights were not initially considered in the discussions of equal rights but evolved thank goodness out of common sense and some version of social efficiency.

us don't understand is that our efforts support few human outcomes that are ever socially or economically equal. People simply experience perpetually imbalanced outcomes throughout all of human history, much like similar natural findings within the greater universe.[25]

But we are able to learn something from all of this, albeit slowly, thereby giving human history a directional arrow that seems productive and purposeful. Some of us learn important, productive values early in life, while others take longer, as I have seen within my own frame of reference. This book helps explains a few of these fundamental imbalances within human society, and I sincerely hope my children are able to eventually learn something as disproportionately valuable as I have.

I originally envisioned this literary effort as a series of small life lessons discussed informally during family meals as a way to selfishly teach my own growing children more about their world outside our small family and outside our typical American community.[26]

Through my own efforts and the contributions of several others, I eventually learned that a few interrelated ideas were more generally applicable to everyone, and I therefore felt compelled to share this valuable knowledge with more than just my children. Thomas Jefferson, a scholar who has greatly influenced me, once said, "Knowledge is power,"[27] and to this I would simply add, "Knowledge should be available to everyone," particularly when it can lead to greater happiness for all!

---

[25] I prefer to think the universe is infinitely fractal and continually expanding (not like the current concept of an expansion to be followed by eventual collapse). I believe areas of continued energy-matter efficiency (in certain geographies) can continue to evolve perpetually if the inputs are correct and thereby defy certain thermodynamic principles such as the law of entropy.

[26] I have been blessed to be able to have dinner with all my children and family for the majority of their childhood lives.

[27] The full quote from Thomas Jefferson is "Perceive the important truths, that knowledge is power, that knowledge is safety, and that knowledge is happiness" (1817). Jefferson was the creative genius behind my institution of higher education, University of Virginia, as well as being a great leader of our country. Education is exactly this—learning from others, adopting what works for you, and perhaps even teaching that to the next generation. I cannot overemphasize the value of a quality education. We need to work on improving opportunity for education within America to improve all the outcomes I am going to discuss.

I did not invent these powerful ideas. They come from a few enlightened thinkers who spent a large and imbalanced portion of their lives communicating these vital ideas to others. For me to summarize their lifework in a few chapters seems almost insulting to their legacy, but in all honesty, it is the most efficient way. Most of us are fortunate if we can remember even a single part of a book title, not to mention its entire name or a small fraction of the content. The best we can do is to get the basic concepts and hope to hold on to these by reinforcement with time and repetition[28] or through some mnemonic tool. This is in part the very idea I am trying to teach my own kids with the idea of *getting more for less*—you don't need to learn everything, just a few important things.

Efficiency has everything to do with how our brain functions (as a naturally skewed neuronal network to process stimuli and rank outcomes by priority),[29] how we learn and even forget information,[30] how we perceive others, and how we thereby form memories through experiences that shape and modify our behavior. I chose the easy-to-remember nursery rhyme (which I interestingly remembered as "Doctor, Lawyer, Merchant, Chief" and not as "Tinker Tailor") both to help my children and to help me!

The rhyme, which I recalled through no mere neurocognitive coincidence, helps me to associate a few cultural ideas, like what is popular within America, with our choices. The playful counting rhyme is an example of what is known as a meme.[31] Memes are a powerful force in our culture that

---

[28] Learning curves are similar imbalanced distributions. Memory can be improved through tools like repetition and association (like mnemonics, which the Greeks and later the Romans emphasized in public speaking and thus in communication).

[29] Steven's law is a psychophysical power law that briefly states this.

[30] Learning curves (and forgetting curves) and all these ideas relate to these same mechanically efficient processes, although they are exponential in this case. We also tend to dichotomize everything into good or bad, rich or poor, *more or less*.

[31] This was a new word for me. Richard Dawkins in *The Selfish Gene* coined the word *meme* to describe an idea that gets copied or mimicked in society and then efficiently propagated (to Dawkins the human is a vehicle to efficiently carry the replicators, our DNA). A meme, as Dawkins defines it, develops its own form that continues to be reinforced over time whether you are there to repeat it or not. A nursery rhyme is a good example, as is a jingle or a slogan like Just Do It. Other examples are "The rich get richer, and the poor get poorer" or a symbol like the American flag. The point is ideas can become their own culture (triggering judgment or ideas), which is a big part of what we have experienced in America with these imbalances.

can invisibly guide our choices and our cultural behaviors. I remembered the rhyme from my childhood, and I cannot exclude the possibility that it played some invisible role in my own life choices and the resulting imbalanced outcomes I have experienced. The rhyme therefore offers a meaningful tool to examine a subtle link between our life decisions in general and their skewed outcomes. "Tinker Tailor" reveals what we consistently value through popularity in our culture (doctors, lawyers, merchants, and chiefs consistently do disproportionately well in America, for example).

My children will learn to ask their own questions and they will face a few vital decisions during their first twenty years, including practical matters such as occupational planning or the extent of their education. But also important are the decisions they make in the core values they integrate from our family of origin, the peers they select and identify with, and the behaviors they choose to adopt from my wife and me and from the larger culture. Little do they realize society is grading them on these multiple-choice questions their entire lives, just as their teachers did in school. As teenagers and later as adults, these choices can become healthy or risky (e.g., impulsive behavior or substance abuse) and productive or wasted, which has everything to do with the physiologically disproportionate growth of the human brain and its temporally dependent hardwiring.[32] All these outcomes follow similar physical laws that relate to the ideas I wish to now enumerate for them while it is disproportionately relevant to their developing brains.

Relatively speaking, I have not learned a lot from my personal life experience, because honestly most of us don't learn much at all, and I am certainly not profoundly different. I have neither invented anything new nor discovered a cure for something as ubiquitous as cancer,[33] but I am not done! As Michelangelo said while still in his hardworking and highly productive

---

32 OECD, *Understanding the Brain: The Birth of a Learning Science* (Paris: OECD Publishing, 2007), 166–67. I would emphasize that even though the brain may be *more or less* sensitive at critical periods (and often follows power law relationships), it is thought of as plastic and adaptable, and *it is never too late to learn.*

33 Cancer and heart disease are two illnesses that account for the majority of deaths in America. Do you think this is a coincidence? I've noticed this since I treat a disproportionate amount of a few diseases every day, and they seem quite predictable in their behavior. There are no coincidences,

twenties, "I am still learning."[34] Most of what we each achieve and learn in life is relative, either to our own expectations, to our small communities, or to whatever frame of reference we feel compelled to compare ourselves. For me, I have spent a disproportionate amount of my life doing things to help improve the health of others (I treat cancers daily, and I help with both quality and quantity of life), and this has been profoundly rewarding, as I feel I have returned to God (my absolute frame of reference) most of the gifts He chose to bestow upon me. Others may have different measures they consider as important to them, but the ideas are still the same. We all have unique gifts, and it is through a better understanding of our choices and consequences that we learn to develop them productively.

At the end of our lives, and with the benefit of hindsight, what most of us discover is that much of our life was a filler of space and time while only a small part seems truly important to us. More importantly, if you have accomplished your ultimate purpose for being here in the time and within the space you currently inhabit, you will have learned something incredibly valuable to pass on to others (such as your children in the case of family or to society in the case of advancements in science or charity), which hopefully also improves our world.[35] Learning these ideas and finding out what is important to you and pursuing it passionately using these few principles can be disproportionately rewarding if you get it early, as I was fortunate to learn.

In my early childhood school years, I remember hearing that we are able to remember only 10–20 percent of what we learn. This numerical

---

my dear reader. See image 1A in the appendix for a graphical representation that introduces you to a power-law distribution, in this case a rank distribution of diseases.

[34] Michelangelo is credited for this, and like many great men, he was always learning throughout his life. *Wikipedia*, s.v. "*David* (Michelangelo)," last modified June 4, 2014, http://en.wikipedia.org/wiki/David (Michelangelo).

[35] If you are lucky, you will touch the lives of a few in a profound way. Similarly you can usefully impact the lives of many others, although likely in a lesser degree. It is not a contest, but the sooner you realize this, the happier you will be and the longer that happiness will last. Life is mostly therefore about relationships, and how you choose to do this is your final report card and depends entirely on your own frame of reference. For me this is Jesus Christ and our Lord God acting through me as a vehicle to help others. For others, another absolute frame of reference will be the judge of your ability to get this idea of doing *more* (for Him) with *less* effort (from you).

predictability fascinated me as a child, and I never really considered it again until I heard this numerical fact repeated upon matriculation to medical school. I can vouch for the fact that I have forgotten disproportionately more than I have ever learned, and this truism is quite accurate relatively speaking! That statistic may at first sound alarming to you, but it is a simple reality. We are constantly bombarded by information, and humans have naturally evolved in such a way that allows us to selectively filter and process what is most valuable to us through feedback and various forms of reinforcement while ignoring the rest.

This rule of thumb of how little we actually are able to learn[36] is a relative truth.[37] Most (at least 80–90 percent) of what we initially learn we forget.[38] I am not any more or less smart than you in the bigger-picture frame of reference. I simply have a different skill set that I have efficiently developed. Roughly 10–20 percent of what either of us learns we retain if we are lucky, especially if it is reinforced with time and repetition. This is why people have to sometimes reference prior generations to rediscover similar information about the things we think are unique to our current human experience (after all, there is rarely a need to reinvent the wheel). As time progresses and knowledge continues to exponentially grow, this disproportional relationship between what we learn and what we forget remains a relative constant and disproportionately so.[39] The imbalance in

---

[36] I used a pun here for a reason. Thumbs and fingers (digits in general) are invisibly guided in their growth by a power-law principle known as *allometric scaling*, which has to do with efficiencies in maximizing physiological surface areas and volumes against energy expenditures. With *allometric scaling* certain organs grow as a power of our body mass. The thumb is then number one in rank of importance for our digits as it allows prehensile grip and other complex (efficient) abilities that allow us to shape and modify our environment to better favor us. Not many creatures enjoy this efficiency or ability, and it is one feature that helps favor our continued existence.

[37] I say *relative* truth because I am not sure there are any absolute truths. Everything is relative to something else. Good is relative to bad, love to hate, heaven to purgatory. There is always an antithesis making it truly hard to define any absolute truth. I could philosophize these outcomes we measure are just that—man's struggle between two opposing forces in everything we choose to do, value, or experience—and the measures we observe are analogous to our population's pulse or the heartbeat of the collective culture.

[38] These are known as *learning curves* and *forgetting curves* and are similar in principle.

[39] At some point our cerebral cortex will become disproportionately larger so we can process all this information (believed to be an evolutionary development), but for now, I think we are in no threat of getting bigger heads.

what we are able to achieve in our memory capacity versus what we wish to achieve is not something seen only with learning or in our basic educational outcomes. It is more simply a repeating theme throughout most of our human existence.[40]

A few obvious examples of these repeating themes that involve predictable imbalances—including how we inefficiently learn—are more commonly visible to American popular[41] culture. For example, most people in the United States (approximately 70–80 percent) *earn little* income and *learn little* (i.e., achieve minimal levels of education), while a few (20–30 percent) earn a lot more of both. I do not believe certain people are inherently better than any others. However, we clearly disproportionately reward *the vital few* (e.g., top 20–30 percent) based on certain critical levels of achievement and educational attainment within our society. Most of these rewards follow rational supply-and-demand economics (also somewhat skewed), while some involve something more complex and less rational, such as cultural popularity that reflects our societal values. Athletes and entertainers are modern examples of this. While our social systems are relatively complex and efficient, we as individuals are not.[42] American behavior is quite repetitious and predictable.

For decades, experts have reiterated the correlation of higher education with more income, yet despite this cultural awareness something has kept Americans from universally adopting these repeating core values into our frame of reference. Many children within our nation, including mine, would rather aspire to be superstar athletes, supermodels, or popular

---

[40] Life plotted as a curve over time is decaying much in the same way disproportionately from the time we are born to the time we die. This becomes more perceptible as you get older and realize that time follows these same rules of imbalance, which is why I want to teach you these concepts early in life.

[41] By *popular* I mean the definition of what is the most common outcome (i.e., not what we desire but what we get). A few people get more, and more people get less. It is therefore natural for the more common group that gets less to always want more. Most people perpetually want what they don't have because they don't know how to get it.

[42] What I imply here is that we make changes and equilibrium is reestablished by the efficiency of the natural system, such as with *supply and demand*. We, however, think we know better or desire more without making the necessary changes to get there. There is a mental disconnect between what we desire and what we are able to do to achieve it. This represents the inefficiency of the individual.

entertainers since these careers represent the truly *elite* within American pop[43] culture. Yet the reality is that these are not popular by population outcome results. And not only are they unpopular by outcome result, but they are so incredibly rare to almost seem to occur by random chance (e.g., my kids are as likely to win the lottery as to become a superstar athlete).[44]

Most of America's children are unable to critically think or plan to do what is more commonly needed within our culture (e.g., getting better educated or finding an appropriate career match for their skills).[45] Meanwhile they are busy dreaming about and spending a disproportionate amount of their time watching super-rare millionaire and billionaire[46] entertainers and athletes. According to Thomas L. Friedman, "American youth are becoming more interested in being entertained than they are in doing the hard yards to earn university degrees."[47]

The reasons for American children's inabilities are vast and complex and for some include social as well as economic barriers. For example, socially, lack of education within a family household can become self-fulfilling to children; economically, lack of opportunity tends to lead to less income. For some, the barriers are physiological and come later. For example, the brain develops more slowly for boys than girls, and sometimes it is not fully mature until the third decade, by which time some critical opportunities have passed. For others, the barriers are more simply a lack of effort, less ambition, cultural (like a growing apathy for religious values), or an inability to understand the nature of just a few vital few choices and what future

---

[43] Here I infer what is popular in *desire* or *appeal.*

[44] I believe that winning the lottery and becoming a professional athlete both rely on luck in large part, since many outstanding athletes don't succeed professionally for other reasons besides just skill.

[45] The probability of winning the lottery is about 1 in 0.5 million to 1 in 276 million for the very popular Powerball. The probability of being a superstar athlete, supermodel, hit musician, or movie star is about the same. The probability of being a doctor is just over 1 in 200 (0.6 percent), and the probability of getting a bachelor's or higher is roughly 1 in 4 to 1 in 3 (approximately 28–30 percent) depending on where you live. I am not a gambler by profession, but I like the odds of education. What is popular with our kids is the freakishly rare stuff, while what is more likely to work in our society and give disproportionately more rewards is not popular.

[46] The billionaire is the modern version of my childhood reference of the millionaire.

[47] Thomas L. Friedman, *The World is Flat: A Brief History of the Twenty-First Century* (New York: Farrar, Straus and Giroux, 2005). Quote is from "Must Read Summaries" in part 4, "A Framework for Moving Forward from Here" (Kindle e-book).

consequences they imply through natural mathematical law and scientific principles.

Some of these outcomes also stem from the inefficient communication of this information by our government and educational elite as well as the inefficient processing of this data by the intended recipients.[48] As an example, consider the input-outcome relationship of cigarette smoking and lung cancer, which I experience on a daily basis as a cancer physician. Lung cancer in 85–90 percent of instances is caused either directly or indirectly by repeated predictable inputs—cigarettes and their many known carcinogens.[49] A few deadly causes (choices) are implicated and responsible for the majority of lung cancers in our country (not to mention other diseases), and yet we continue to choose to smoke despite federally mandated campaigns to make this information publicly widespread. The *why* in this case has a lot to do with the complex nature of human behavior and the link between these carcinogens and powerfully addicting substances such as nicotine. Humans have successfully created and marketed a vehicle of incredible efficiency that can deliver a substance widely used (similar to other drugs of abuse) as a carrier of these carcinogens to their final destination in your lungs and elsewhere.

And furthermore, despite our awareness of tobacco's dangers to our health culturally and armed with this knowledge as individuals, we continue to promote it as an industry (for the sake of wealth), and we make the daily choice to value it within our personal lives. Where I live, by no coincidence, is among the highest user population of tobacco within America. As you might imagine, I therefore see a disproportionate number of Americans

---

[48] As an example, to properly analyze income imbalance within America, you would need to stratify by many variables—not just gender, age, ethnic background, geography, or education. You would need to compare different results by taking these into consideration in your analysis as multiple co-variables to see which ones are statistically significant.

[49] This is from CDC information and the National Institutes of Health. US Department of Health and Human Services, *The Health Consequences of Smoking: A Report of the Surgeon General—Smoking among Adults in the United States: Cancer* (Atlanta, GA: US Department of Health and Human Services, Centers for Disease Control and Prevention, National Center for Chronic Disease Prevention and Health Promotion, Office on Smoking and Health, 2004), www.cdc.gov/cancer/dcpc/prevention/other.htm.

that suffer direct health consequences and who ultimately die[50] as a result of this mostly preventable disease that comes from a simple human choice.

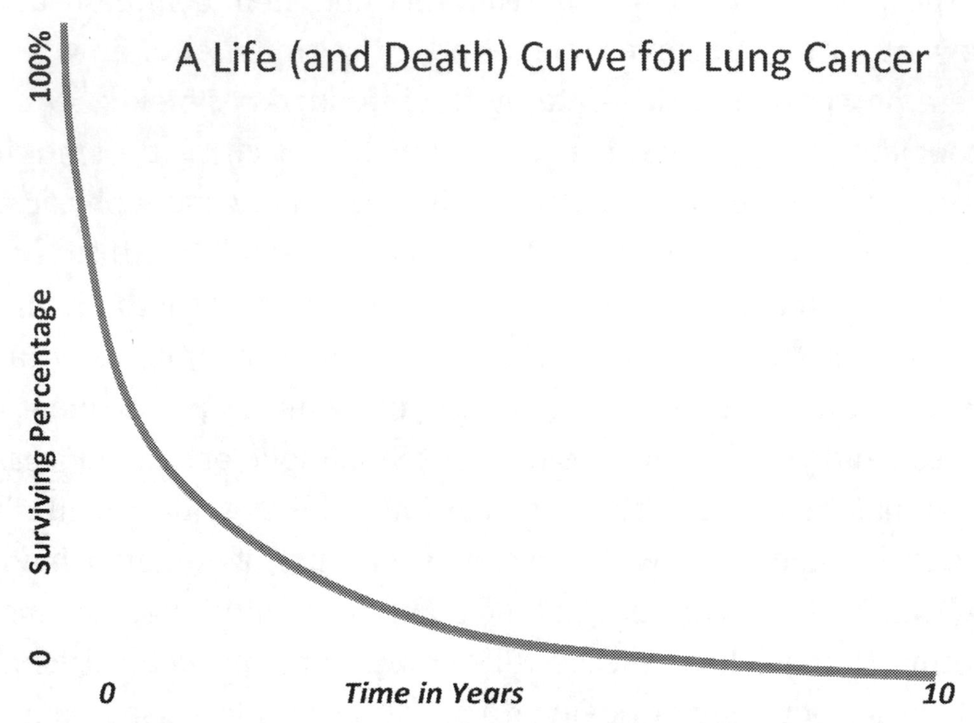

Most deaths occur in the first few years. The % of survivors drop precipitously with time

The image shown represents a typical *(life) survival curve* for people diagnosed with stage III and IV lung cancer. Certain cancers, such as lung cancer, that disproportionately present in later stages (roughly 80 percent are in later stages) are efficient killers of humans. However, many cancers are mostly preventable by learning to modify our human choices.[51] It is the

---

50 West Virginia has the highest incidence of tobacco per capita (116.3 men and 71.3 women per 100,000 population was the highest for 2003–2007 for men and second highest for women next to Kentucky) despite the fact we are not a big grower of tobacco in comparison to neighboring states like Virginia or North Carolina. West Virginia also has one of the highest death rates in America for lung cancer (for all cancers collectively we are number one). American Cancer Society, *Cancer Facts and Figures 2011* (Atlanta: American Cancer Society, 2011). Also, as a side note, the cure rate for lung cancer and all others drops by almost half for those who have attained high school education or less versus those with more. More education means fewer deaths from cancer as well as many other diseases. Ibid., 25. You can see why I have a keen interest in both ideas. Education is really a more efficient tool for treating this disease.

51 It is guesstimated that over 50 percent of cancers come from our diet, smoking, alcohol, lack of exercise, and other habits that involve human choice (certain viruses, for example, like HPV, are

very nature of this repeating skewed curve that made me first aware of the principles of imbalance throughout our lives and the concept of choosing efficiency not just in our health but also in other outcomes that are similarly dependent on our free will.

I find it curiously coincidental that survival curves for a disease like cancer can mirror the mechanics of income distribution within America, and yes, there is certainly some link between our health and our wealth. But there is much more to this link than money and better health outcomes. There is a subtler link that I am interested in, and it has to do with these recurring imbalances and what they imply about human choice and consequence (or, questions and answers) and what Americans *choose* to value.

I believe there is a fundamental mismatch between many of the values we choose to adopt within America (both for our desires and our expectations) and our final achievements. What we want and what we are able to get are diametrically opposed variables that are related by a few important principles that I have uncovered and that I will use as the framework for this book.

Similar Imbalances of Wealth and Health Because of Human Choices and a Vital Few Rules

Income distribution to Americans is a topic ripe with data and is interesting to more people than health outcomes, and so I use it as a common model in this book to highlight these repeating principles of imbalance. It is not that I am fixated on money more than health (no more or less than our proverbial Daisy), but the distribution of money within America appropriately and efficiently provides a wealth of knowledge. It is no mere coincidence that money is linked to the same ideas of health imbalances that I observe disproportionately every day in my career as the *doctor* as well, and I hope to show you the nature of this *power*ful link throughout this book.

linked to cancers). I would argue that it is much harder to change human choice than to regulate our industries. The fact we let a substance of such affinity to our brain pleasure centers like nicotine be linked to a product (that is freely marketed in our society) that is carcinogenic in multiple ways when combusted astounds even me. I would also point out that it takes significant time (decades) to see the effects quitting this habit has on the risk of cancer.

Why the interest for me in health and wealth and in rhymes and reasons? I only became personally interested over the last few years in critically analyzing the distribution of wealth (and likewise health) within America.[52] I have mostly had plenty of money for years, and so in all honesty I really never gave its allocation within America a second thought—it simply seemed to naturally follow the consequences of my own choices and actions.[53] But more recently I became interested in the spread of income for several converging reasons.

For one, society seems to have become frustrated with the perceived growing trends for income dispersion within America, which according to experts appears oddly out of whack. Our current president, Barack Obama, recently emphasized the fact, "Income disparity is one of the greatest challenges currently facing our children's generation."[54] I therefore felt a pressing need to better understand the dynamics of the seemingly timeless relationship between Daisy's proverbial *rich man* and *poor man* and even the *beggar man* and *thief*—if not for American society then for my own peace of mind.

For a second and a more personal reason, my teenage children are nearing adulthood, and they will have to provide for themselves soon. So practically speaking, I needed to brush up my knowledge base in order to help guide them to make appropriate career choices and efficient life choices.[55] While I quickly confirmed that income is mysteriously imbalanced (as are careers and American education[56]), I later discovered there is much more to this

---

52 One of the first things I noted was the skew of health outcomes and wealth relative to our state of West Virginia. We suffer from the worst outcomes for cancer deaths in the country as well as the poorest income levels in the nation. It is no coincidence.

53 This is an important point. My actions have been consistently valued within society, each disproportionately so such that the product of many such skewed outcomes are even more skewed (educational achievement, occupation, work ethic, etc.).

54 This statement was from December 2013 from a speech by Obama on income inequality. Jim Kuhnhenn, "Obama: Income Inequality Is 'Defining Challenge of Our Time,'" *Huffington Post*, December 4, 2013, last modified December 5, 2013, http://www.huffingtonpost.com/2013/12/04/obama-income-inequality_n_4384843.html.

55 I could likewise suggest marital choices, which is another reason I chose this interesting nursery rhyme. In fact, this rhyme has many of the key elements constituting my values—materialism, religion, marriage, careers, family, etc.—providing a great starting point for discussion of these concepts.

56 For example, in one recent survey by the US Census Bureau, West Virginia had the twelfth-highest rate of state uneducation (in people over twenty-six years of age) below ninth grade (6.9 percent),

story of imbalance than a simple randomly unequal distribution of money within American culture.

My children have unconsciously experienced this imbalance on a personal level for years, since their dad earns more than 99 percent of the people in our community, our state, our nation, and in the world. Many of us, including my kids, *are* acutely aware that some monetary imbalance exists in basic concept, but we are not cognizant of the true nature or the extent of the imbalance or even how long it has been present within America. For my kids to understand why these unequal outcomes occur or to even care enough to see a correlation with their future lives, I had to learn more about income distribution myself. Even I, who consider myself reasonably well educated, learned something powerful from these findings of (income) imbalance, including how it can metastasize to negatively affect our objective health and our subjective well-being as a nation, which is why I was compelled to share it with others as well.

In addition to becoming a doctor in America as my own fortunate choice of career, I am also proud to be a Christian (a Methodist), a father, a husband, and a family man. The joy and happiness from these vital few choices will far outweigh many other outcomes I receive here on earth. My four children are now old enough to begin to independently make their own decisions about whatever they want to get out of life, much in the same way as the symbolic Daisy from the nursery rhyme. Like most parents, I simply want to help my kids learn to make their own appropriate choices and to be happy—whatever they choose to do, provided it does not harm them or others. This can be the case for their occupational choice, where they choose to live, and with whom they decide to spend that time—all examples we see recited in a simple rhyme by a nursery school–aged child.

In the ideal world, I would furthermore suggest that what our children choose to do with their lives should efficiently match their unique, God-given

---

the eighth-highest failure to complete high school (18.5 percent), the forty-third-lowest rate of high school or higher education (81.4 percent), the lowest level of attainment of associate degrees or better (22.8 percent), the lowest of bachelor's (17 percent), and next to the lowest for graduate degrees (6.6 percent). US Census Bureau, *Fact Sheet: 2006–2008 American Community Survey 3-Year Estimates.*

talents with their passions and desires, while considering the greater culture, incorporating meaningful core values, and preferably helping make society a better place. These are what I refer to as a *win-win* outcomes—they benefit our children *and* help make America a better place, providing a perpetually efficient outcome for them and future generations.

Most importantly—and this is where I find it hardest as a parent—my children's choices have to be their own and not necessarily according to my will. If we as parents choose to measure our children's abilities in relation to others, we must be careful, as I have learned the hard way. Our children have unique gifts and attributes that make them each special. Famous German-born scientist Einstein (who chose to become a dual Swiss American) intelligently stated the obvious, "Everybody is a genius. But if you judge a fish by its ability to climb a tree, it will live its whole life believing that it is stupid."[57]

What I believe I have unmasked in my research to help each of my children is an invisible link between basic human behavioral *choice* and *consequence*, which, by no coincidence, follows the same nonlinear imbalance as American income.[58] Even childhood happiness, as it has been measured domestically and internationally, is a natural consequence of human behavior and is repeatedly skewed in similar ways, as I have learned from my own four children, and relates to these same ideas![59]

---

[57] "Albert Einstein Quotes," *Albert Einstein Site Online*, 2012, http://www.alberteinsteinsite.com/quotes/einsteinquotes.html. Albert Einstein was a physicist, which qualifies in my book as a tinker. His nationality changed quite often when he was young due to cultural biases, and he emigrated from Germany (as did many scientists from that time) due to the Nazi regime's persecution of fellow Jews. He finally settled on a dual Swiss American citizenship. Many scientists today still gravitate to the United States for different reasons, including the opportunity to pursue work at quality institutions of higher learning (e.g. Princeton and Harvard), creating a brain drain on the rest of the world, which has and continues to disproportionately help America. Many of these concepts develop similarly as human attachments to others that are self-similar, creating a growth process that becomes a power law.

[58] Pareto, whom I describe in part 1 of this book, used the term *vincolo* or "bond" to relate the two forces or factors that link together or correlate. Pareto, *The Mind and Society*, 1:67. This is the first example of a power law.

[59] While three of my kids (75 percent) have adopted our family values, one (25 percent) is an exception. There is always an exception to everything, and there is often a similar pattern to this simple mathematical relationship. So my children will naturally each serve as great models for

Others, for example, have described a reciprocal relationship between the well-being of a nation (and its children) and the perception of income imbalance.[60] While happiness may not follow the same exact mathematical relationship as some of these other principles that I am going to teach you, it does follow the general *principle of getting more for less. More* happiness and greater well-being are seen with *less* perceived income disparity. Although this inverse link between income disparity and well-being was a surprise even to me (I assumed it was simply poverty alone that correlated with misery), it does not come as a complete shock, especially as I look around at the growing unhappiness among youth within American culture.

Currently, America has one of the highest income disparities in the modern world of industrialized nations, and we also coincidentally have the worst well-being in our children! This is evident in (1) our children's mental health as evidenced with drug use, suicides, homicides, violent crimes, and prison terms and (2) their physical health with measures of infant mortality, life expectancy, teenage pregnancies, childhood obesity, and chronic diseases, as well as many other ways that we are only now beginning to understand.[61]

What I find most telling about this information, from extensive research on this particular subject of income disparity, is the lack of a correlation between simple poverty and these other measures of our well-being. Being a *poor man* is not enough to explain the misery within our society; rather it is the wide skew of our distribution of income (i.e., income inequality itself) that is adversely linked to these poor social outcomes.[62] It is therefore not as easy as simply blaming all the bad outcomes in society on the poor within our country. In fact the relationship between income per capita and

---

the principles that I wish to teach (representing both good and bad outcomes) in my discussions throughout.

[60] Richard Wilkinson and Kate Pickett, *The Spirit Level: Why More Equal Societies Almost Always Do Better* (London: Allen Lane, 2009).

[61] Ibid.

[62] I personally believe all these models measure the same thing. It is not simply income disparity that creates these outcomes. Human behavior creates all the same outcomes, and income disparity is just one model of it, so too are educational, religious, occupational, marital, and so on disparities. While fixing income disparity may help, it is not the end-all solution to a more general behavioral pattern that affects more than income.

well-being does not correlate like income inequality does. Likewise, it is not as simple as blaming the *rich man* (the top 1 percent) for the other side of the issue. The issue is more simply how we perceptibly compare and rank ourselves to others within our popular culture and how we choose to treat each other as fellow human beings, as subtly communicated by the simple nursery rhyme "Tinker Tailor."

What is to blame is the increasing relative division between these two extremes that we have allowed our own society to propagate. We've allowed this for many reasons, and a vital few are worth our efforts to learn. I hope to help reshape these cultural values for our children by showing them both a rhyme and a reason for the unequal outcomes within American popular culture.

The link with unhappiness may be more geographically obvious to some like myself, since the area where I originally chose to live is repeatedly one of the poorest states within the United States. However, it is, in fact, more often the relative imbalance of income that is more correlated geographically with repeatedly bad societal outcomes. Yes, poverty matters wherever you live, but what truly shapes our well-being and our choices (including violence, crime, depression, drug abuse, suicide, etc.) is the relative perception of how poor we think we are as a community and how rich we think we are as a nation, as compared to others around us.

This repeating thought pattern is prevalent everywhere within America, although the negative outcomes cluster in certain geographic hot spots, much like similar geographic groupings of naturally violent volcanoes or skewed weather patterns.[63] Learning this is important if you wish to be able to choose where you live and have it remain a hospitable geography.

In the bigger picture, understanding how humans perceive ourselves relative to each other through rank and comparison[64] and learning to

---

63 For example, crime is disproportionately high in certain geographies, such as Detroit, Michigan. My small West Virginia community has had an influx of drug dealers from Detroit that has permeated like a virus over the past decade, and that has negatively impacted my family and state. Viruses, by no coincidence, just like certain ideas (in this case bad ones) spread by power laws. Cancer metastasizes quite similarly as well, especially when our defenses are compromised or are otherwise inefficient. Earthquakes and volcanoes follow similar geographic clustering in distributions for a reason.

64 Rousseau, a thinker much smarter than I, said this same thing. I found his discourse after I had composed my own ideas, but it seems too coincidental to ignore: "In fact, the real source of all of

treat each other more lovingly and equitably is therefore crucial, as it disproportionately shapes our sense of well-being everywhere. This is why, even though Americans live a higher standard than almost every other country in the world (when you look at income per capita), we are still disproportionately unhappy and unhealthy. This unhappiness has everything to do with relativity and how we perceive ourselves compared to one another and how we treat one another. This is what I therefore became interested in quantitating for our children and for myself as an American.

As a parent who is concerned about the happiness, health and general well-being of my children and their future within our great country, these ideas should be alarming. I have experienced some of the negative effects firsthand with one of my children, so I can assure you it is a real phenomenon, and for me it was truly a wake-up call! America currently has the *worst* well-being scores (i.e., our kids are the most miserable) internationally among children![65] So what does this mean to us as a nation?

Most of us are probably too provincial to worry about how the rest of the world *feels* or how income disparity affects another country's well-being.[66] We are usually more concerned with local and regional problems. So if you don't particularly care about America in comparison to the rest of the world, then pull your head out of the sand and look around yourself and even closer to home! The same fractal relationship is seen within all of our states—some more than others—and in the same skewed way as can be observed worldwide.

If you naively *assume*[67] that you are immune to the effects of disparity because you think as I did that you can shelter your kids from all this, think again! There is no utopian microenvironment within America. We

---

these differences is that the savage lives within himself, whereas the social man, constantly outside himself, knows only how to live in the opinion of others; and it is, if I may say so, merely from their judgment of him that he derives the consciousness of his own existence." Jean-Jacques Rousseau, *The Social Contract and Discourse on the Origin of Inequality*, ed. Lester G. Crocker (New York: Washington Square Press, 1967), 245.

65 Wilkinson and Pickett, *The Spirit Level.*

66 Zipf, who describes this idea of imbalance further in his book, uses this analogy: "A boil on your neck will worry you more than a famine in China," unless of course you are close to China, in which case the boil becomes less relevant when compared to starving to death!

67 You know what they say when you assume? You make an *ass* of *u* and *me.*

are all interconnected in our imperfect and imbalanced ways, and I have realized this myself now more than ever! My own son, despite a privileged social and economic childhood, chose to identify with peers who do not value educational achievement (relative to their immediate local culture). He chose a teenage life of repeated drug abuse and linked poor outcomes. West Virginia, his home during childhood in fact has an unusually skewed incidence of drug abuse in teens, which also correlates with income disparity and the viral-like spread of many of these undesirable behaviors within our society. This drug abuse is ironically made more efficient by the medical industry through prescription drugs.[68]

I have found that these outcomes,[69] by no coincidence, relate to the cultural imbalances seen geographically with measures like income. My son's example simply exemplifies how these imbalances can infiltrate any geography and negatively affect our kids' mental health, similar to an aggressive visceral malignancy, such as lung cancer.[70]

After further exploring several of these repeating imbalanced outcomes in my own life, particularly as they pertained to my children, I was able to distill the relevant information into a few simple principles that I believe are vital to our understanding.[71]

---

[68] The number one category of prescription medicines within America is pain medications (see graph 2A in the appendix). Number two is anti-cholesterol medications to counter our dietary choices, and number three is anti-depressants as a group, since we are so miserable as a country! My point is that we physicians and the pharmaceutical industries enable the problems as well as trying to fix them since we all have vested interests in what we do and that these tools that we use work efficiently by design. However, we are creating a society that learns to need (i.e., become addicted to) pills for everything.

[69] In *The Spirit Level*, Wilkinson and Pickett show that every one of these outcomes correlates with income disparity (both at the state level within America and on an international level as well). I personally believe they all represent growing disparities in our basic values—with income simply one model, just as health, education, religion, etc., are models.

[70] Malignant cancers can infiltrate our bodies efficiently (for them) in ways that mathematically follow these same skewed principles, which is why time is of the essence in treating these conditions. Once cancer growth reaches a certain threshold, cures become disproportionately less likely. In my line of work, survival curves for very efficient cancers (e.g., lung cancer is an efficient killer with low survival rates) look identical to these imbalanced curves of income, education, occupations, marriage, etc. This similarity between survival or life expectancy curves for diseases and the model of income imbalance and human behavior struck me as too coincidental to ignore.

[71] As an example of my own desire to be more efficient and practice what I preach, the first draft of this book had over one thousand pages and over one hundred graphs. With a lot of effort, I distilled

I believe a few principles of human behavior describe the mechanics of most of life's repeating disparity (including most inequalities within America), and I wish to share these rules with my three American girls, using them (and Daisy from the rhyme) as models. I believe a fourth idea, which I included in my book title, *Getting More for Less,* efficiently embodies them all as a single meme, from which I hope to impart invaluable skills to my son, which could disproportionately affect his future as well.

The three simple principles I explore in this book are socially the equivalent of what gravity is to the physical sciences,[72] and if you think gravity is important, you will appreciate the nature of these forces of attraction as well.[73] They can similarly help you efficiently attract *more for less.* If you are able to learn these three vital principles, you will discover what is relevant within your own life (a few simple core values become your frame of reference, for example) and what is trivial (the rest). Learning a few *power-law principles* can empower you to change the world if you look for them within human nature and social nurture.

In part 1, I will discuss the (Pareto) *principle of imbalance,* as it relates to *money.* This principle describes a particular perpetual imbalance for the distribution of money that has not only been noted by President Obama (the *chief*) in twenty-first-century America but also described originally by Italian engineer Vilfredo Pareto over one hundred years ago. This principle relates to the nature of income distributions but also applies conceptually to much more than just an imbalance of American money, as I hope to show you.

In part 2, I will discuss the (Zipf) *principle of least effort,* which describes a related imbalance between our choices and their consequences—for example, where we choose to live, whom we marry, what newspapers (or

---

it to less than 20 percent of the original amount in order to save you time.

[72] Gravity is an inverse square (a power law) that accounts for the unusual imbalance in the universe that in fact favors our existence. These same rules of physics account for the complex interactions of all matter and energy in the universe, and I propose these social laws are manifestations of the same interactions—in this case many beings of collective matter/energy interacting with similar forces of nature acting on them.

[73] I suggest here a relation to something known as the *law of attraction,* which I heard about through Rhonda Byrne's great book *The Secret* (New York: Atria, 2006) and video. Since watching Byrne's inspiring video probably fifty times since 2007, I have been unconsciously inspired to enumerate her ideas.

Internet search engines) become popular, what clothes are in vogue, which words we choose to use in our speech and how often we use them, and even the mechanics of how we dream! George Zipf, a Harvard educator, introduced this social concept over sixty years ago, and it relates directly to the mechanism of these complex relationships. I will emphasize the role of our efforts and how our choices consistently lead to disproportional outcomes, whether we are aware of this principle or not.

In part 3, I explore a concept known as the (Juran) *80–20 principle*, which describes a repeatedly skewed numerical relationship between any two correlated variables that usually involve quality measures (good or bad). This principle has many applications to business, industry, and even to our social and personal lives. Joseph M. Juran, another engineer,[74] pioneered and first applied this tool widely in business following World War II. This principle has real-world applications in just about everything involving *value* and the related concept of *quality.* I place an emphasis on how we can use this principle to reshape our core values through choice and their linked actions within modern-day America.

All three highly educated thinkers developed similar ideas from the same findings of repeating and predictable imbalances that were pervasive during their lifetimes. I believe their findings are intimately related to each other and apply to our modern-day lives and in fact have formed the basis for many great thinkers over the last century. These analytical thinkers all reasoned similarly that life obeys predictable principles—with human behavior being merely one such example. All three men devoted the majority of their years trying to passionately educate others about these powerful principles of imbalance! In fact, your time and how you choose to spend it, whether you realize it or not, is also similarly skewed in its distribution. I will allude briefly to the idea of reshaping your skewed *time curve* later, but trust me when I say that learning this information and a vital few concepts is time well spent if you truly grasp these ideas!

---

[74] If you haven't noticed a coincidence here yet, with engineers and a few other occupations, pay more attention.

In the final part of this book, I show how using these few (three) vitally important rules can help us to reshape the way we all think about what is popular within American culture, how we choose to spend our time, and how we can be more efficient in our efforts as individuals and as a society. While income disparity may be the model I use for America, the applications can be retooled for any aspects of your life where you see outcomes you wish to change. *Getting more for less*, as the final idea in part 4, then embodies all three of these principles and teaches you how to be efficient in linking your choices (e.g., values) with their intended consequences, either as an individual or in a larger population cluster.

President Obama has recently emphasized that "we cannot promise equal outcomes [for things like income], but we can promise equal opportunities."[75] His statement applies to the roles we have as a society to help minimize opportunity-linked disparity related to gender, ethnicity, race,[76] sexual preference, religion, and geography so that more of us are able to achieve the ideal American dream. How we define this ideal is subject to philosophical debate, but I hope to give us all the best tool of all to equalize (I should say *make less unequal*) the playing field for achieving these outcomes. And while I cannot promise everyone, including my own children, equal outcomes based on what they are able to learn from these principles, I can promise the same opportunities to achieve happiness by presenting everything I believe is relevant concerning a few vital social laws that guide our underlying human nature, including how we learn to think and act. The rest is then *more or less* up to you and the power of your free will!

---

[75] This quote was taken from an Obama speech in December 2013 that I watched on TV in which he emphasized this growing problem in our otherwise great nation.

[76] You may notice that I tried to avoid ethnic, race, and sex preferences as variables in the analyses. The reason is that I prefer to focus on other things that we are more likely able to change about ourselves.

PART 1

Pareto and the Principle of Imbalance

"Rich Man, Poor Man, Beggar Man, Thief."

The Principle of Imbalance and My Own Family

Admittedly, I am the *rich man* from the rhyme, but on several occasions in my life, I have been the *poor man* as well. While I have been blessed to never be a *beggar man* (if you don't count my college years) or as desperate as the *thief*,[77] I have been both rich and poor, so I believe I speak from experience when I discuss these various imbalances and how they factor into our lives, socioeconomically and otherwise.

I was born to a socially disgraced, unwed high-school teenager in a rural geographic part of eastern North Carolina almost fifty years ago. Our young American family (of two) lived in relative poverty for several years before eventually achieving social upward mobility by becoming a household family through the combined power of a second marriage, solid values, and higher education.[78]

My first personal awareness of any significant inequality within America came as a child in *middle school* (notice how there is hierarchy built into the educational system early on). I was incredibly fortunate at that time to have the opportunity to switch from a public school education to a private school where my mother taught for a vital few years of my primary education. Our family was unable to afford the private school tuition, but because she was an educator there, I was fortunate and grateful to be afforded entrance. While there I witnessed not only a higher level of general educational quality but more importantly a difference in the value of achievement. More people, as

[77] I do, however, have intimate personal experience with a thief, as one of my children chose this unproductive behavior out of desperation.

[78] My mother eventually graduated high school and achieved a college degree despite having a child at seventeen. She met my future stepfather while in college. I had a Christian upbringing as well.

a relative percentage, seemed to truly enjoy being there. Teachers in that particular environment were clearly more passionate about their efforts as educators for eagerly receptive and ambitious students.[79] In hindsight this was what we refer to as a biased and privileged socioeconomic population, and I was graciously along for the free ride. While this would seem to offer an easy explanation for my future success, life is rarely ever that simple. A few years later, my mother became pregnant once more (this was planned) and chose to quit teaching, forcing me for financial reasons back into a more typical public school. But the lessons I learned during my five-year stay at the private school have remained imprinted on me for life, and for that I am grateful.

I later chose to attain a college degree at a very affordable and respectable public university and did so efficiently (within four years)[80] before accepting my first full-time job. While my starting salary at that time ranked me in the bottom 20 percent of incomes within America,[81] I was nonetheless grateful to become employed and blessed to learn self-sufficiency. I was able to eventually rise in rank socially and economically through the power of my further consistent educational attainment (earning an advanced college degree in medicine), academic achievement (graduating number one in rank from my medical school), and later marriage, similar to my own family of reference.[82]

While I don't believe I am smarter than anyone else or better than others, I do believe my family has enjoyed better opportunities than others and better outcomes thanks in part to our families of origin and what they chose to model in value to us (and thus what became our frames

---

[79] I have tried to duplicate my schooling experience with my four children, giving them choices of either private or public education. They prefer private for some of my same reasons.

[80] I was fortunate to have a great school where I lived (again thanks to my parents) in a city that has one of the top public colleges in the nation—the University of Virginia, which is consistently ranked among the top institutions (private and public) for the nation and is usually top 1 percent for public colleges. College performance is disproportionately ranked with a few colleges consistently scoring well and a lot of colleges performing at lower levels of valued measured outcomes. My college tuition was also less than the annual tuition for the private school that I attended gratis.

[81] I earned $12,000 a year annual salary at the time, which put me in the lowest income quintile in America at the time.

[82] This relates to what I referred to when I said our parents affect us directly or indirectly in so many profound ways that seem so random. We cannot pick our parents, but they sure can disproportionately affect our outcomes by the frame of reference they provide to us.

of reference), coupled with a vital few choices we also made. I am also fortunate to be disproportionately rewarded highly within society based on my achievement and occupational choice, and I am grateful to the degree that I want to help others, including our children, who hopefully care enough to learn from this information.

In the course of my reference lifetime I have personally therefore been through two cycles of poor to rich, with both instances disproportionately assisted by the consistently valued inputs of (1) educational achievement and (2) quality values, as they relate to the individual and to society.

During my own half century of experience of cyclical social mobility, I have realized that our country (indeed, the world) is also socially and economically *imbalanced*, but I did not at first realize to what degree. I don't mean to imply this is a bad thing, just that life seems repetitious and even somehow *predictable* and quantifiable to my analytical way of thinking.

What you now think or thought you knew may be similar to my thoughts. I was like most Americans who significantly underestimate the true extent of our societal inequalities[83] and their recurring nature. In fact, I only became interested in the idea when I discovered I had a problem and after it affected the health of my own wealthy family. We often hear from the media and the governing elite what it is they want to subjectively convey about all these ideas, and we only really tune in when it seems to apply to us. Since I am a bit suspicious of conclusions from the media or the governing elite, I decided to research these ideas objectively on my own. This book is the result.

I first began this project by examining objective US Census data, specifically looking at various and sundry databases for annual incomes in early 2012, as I wanted to learn the natural spread of incomes for our population and the distribution of income for the various occupational decisions facing our children.[84] I decided to review this issue myself to

---

[83] One recent poll revealed Americans thought the top 20 percent owned 59 percent of the nation's wealth, while in reality it is closer to 87 percent of the wealth. Wealth is more imbalanced than income, which is very skewed. Careers are similarly imbalanced as I quickly learned. Michael Norton and Dan Ariely, "Building a Better America—One Wealth Quintile at a Time," *Perspectives on Psychological Science* 6, no. 9 (2011): 11.

[84] I intermix the ideas of occupational incomes and household incomes, but they are both skewed, as I will show you. Occupational incomes are perhaps less skewed and more exponential in their

better understand this particular imbalance as presented in the popular media and how it might change in the future to affect my four growing children, one of whom recently decided postsecondary education was not the right choice for him.

I am fortunate that my job highly reimburses me for my academic degrees, and yes, I am truly grateful to be in the highest income group for all occupations within America. It is ironic, therefore, that I chose to live in a geography consistently ranked among the poorest and the most uneducated within America.[85] This Dickensian paradox seems to have further magnified the disturbing effects of these imbalances to my eyes, giving me quite a unique perspective of my family and even our country. I have truly been blessed as an American, and for most of my family of origin it has been the best of times, but for the majority of the people in our state, my son included, it has simultaneously been the worst of times.[86]

I consider myself a realist, nicely balanced by a wife who is an optimist, and we have concluded that not all our children are likely destined for either of our career paths or either of our unusually high levels of educational achievement. And so I wanted to educate myself about (household) income and occupational data in order to help our children in their own unique career paths and in their related educational efforts. This gave me a selfish motive to find some link between income and the many other variables that correlate besides simple occupational choices. What I learned was even more valuable.

My children are just like most other kids within modern American culture. They like the easy, simple paths to success—as shown by my oldest

behavior, but the general ideas of imbalance are still the same.

85 There are no coincidences, only undiscovered links. West Virginia consistently ranked last (number fifty) by state for levels of educational attainment for the last twenty-plus years, which cannot be a random outcome. There is a concept known as *propinquity* that I will explain later.

86 Charles Dickens wrote often about the plight of the poor and condemned the social hierarchy of society, which he believed industrialization only worsened (I agree). His perspective was no doubt influenced by his early childhood when his family entered a debtor's prison. He was forced to quit his education and work as a menial laborer, putting soles on shoes to provide for his family. When his grandmother later died, his father received an inheritance, which propelled the family out of debt and allowed Dickens to resume his education. This shaped his views on life disproportionately for decades, resulting in his well-known book *A Tale of Two Cities.*

child who decided to simply forego a college education. Easy is more popular within American culture, as exemplified by less educational attainment seen statistically in the general American population. This is, in part, why I picked the title for this book: *Getting More for Less*. This title implies easy, and we all seem to be able to relate to that concept.

My son has in fact *learned* to prefer the easier approach, having seen (1) the way our popular culture rewards musicians who are often uneducated (his preferred form of musical entertainment is gangster rap), (2) how we glorify certain alternative lifestyles (through our less-filtered modern media), and (3) the way we disproportionately reimburse entertainers such as movie stars and super-athletes.[87] I am certain my desire to make his life easier than my own did not help the situation favorably, but as I often reiterate, I am still learning!

My other children similarly have observed how our society rewards outward visual appeal in skewed ways and how sexiness, individualism, athleticism, and consumerism are disproportionately valued while American culture discourages the pursuit of many worthy professions, such as the social service industry (my wife is a social worker), our protective police and firefighting forces, our military (the *soldier* and *sailor*), and our educators. Within each of these, there are exceptions, but in general the rewards are not proportional to the value these occupations merit within our society, but rather to the occupations' popularity and simple math (like supply and demand).

My children witness how disproportionately we reward effortless random events, like winning the state lottery by easily spending a few bucks, while we scarcely reward predictable hard work, such as manual labor or other similar lifelong productive efforts. While there is much to understand about the supply-and-demand economic model of labor

---

[87] I would still point out to my kids that the probability of this happening is basically one in a million for each of these situations. As an example, there are roughly 360 NBA players, and so the odds of being a superstar athlete (in basketball) are roughly 360 out of 316 million or about one in a million. The odds are roughly the same with supermodels. I would rather you plan for something that has a one in five chance of success or even a one in one hundred chance. This magnitude is way more achievable and realistic.

efficiency, the invisible motor[88] that drives these inequalities is most often simple human nature. However, much of the disparity we currently see in America currently—income being merely one example—is because of what we have learned to value within our nation, as well as what we model to our own children through these many machinations of popular culture.

What I hope to teach my kids through my additional research efforts is that a few simple and consistently good values and decisions (like education, marriage, solid core moral and religious values, and yes, even some hard work and achievement) combined with their own unique passions and abilities can help them statistically achieve more happiness and success. And while my children may not realistically ever become athlete superstars, famous rappers, supermodels, or ever win the state lottery[89] on a probability basis, they are capable of making more-efficient choices that will allow them to attain the American dream (e.g., being in the upper middle class[90]), which is much easier (and more probably) to achieve than these other childhood fantasies.

While I do not discourage a super-elite career for anyone, as I am not one to judge our population by any occupation, I do believe a minimum level of

88 This is analogous to Adam Smith's *invisible hand* from his treatise *An Inquiry into the Nature and Causes of the Wealth of Nations* (1776). According to Smith, there are not many differences in our natural abilities, but rather the division of labor comes from opportunities and human choices: "The difference between the most dissimilar characters, between a philosopher and a common street porter, for example, seems to arise not so much from nature, as from habit, custom, and education." Adams Smith, *An Inquiry into the Nature and Causes of the Wealth of Nations*, ed. Jim Manis (Hazleton, PA: Pennsylvania State University, 2005) 20, http://www2.hn.psu.edu/faculty/jmanis/adam-smith/Wealth-Nations.pdf. When it comes to super-athletes and other measures of similar talent, I might beg to differ here, but for most people this is correct.

89 In West Virginia, the lottery is the preferred way out of poverty, with five millionaires created in 2012 from lottery winnings in the state population of 1.8 million. We spent $1.5 billion of our earned income as a state last year to produce only five millionaires, and most of the money of course goes to the state for various allocations and to cover expenses. We are one of the highest states for per capita participation in lottery-style gambling (in 2012 West Virginia spent $810 per capita in non–video lottery terminal sales with only one other state, Rhode Island, spending more). The odds of winning the Powerball are roughly 1 in 276 million, and someone from West Virginia already one this a few years ago. West Virginia residents spend 2.5 percent of their income per capita on lottery tickets in a state already ranked at the bottom within America, which should tell you how desperate we are. On a positive note, the lottery generates huge revenue, which is somewhat efficiently allocated to schools and other programs in the state, such as Promise scholarships, and provides a surplus in the overall state budget. Reference is from LotteryWV2012 Comprehensive Annual Financial Report for the fiscal year ended June 30, 2012 (Mara Pauley, project coordinator). Accessed from wvlottery.com.

90 This is a one in five chance (20 percent) and would be a reasonable goal for anyone with ambition.

educational achievement is a more efficient choice for most American children, and getting this early on will pay off disproportionately in the long term.

The main impetus for my colossal effort was my own deficiency in showing my firstborn child the powerful link between a valuable education beyond high school and his probability of future success and happiness.[91] I emphasized this repeatedly during his early formative years, and I always assumed college was a given choice and outcome. However, our local culture of *under*-education influenced his perceptions through the medium of his peers, and he was *unable* to learn the importance of higher education. Instead of blaming him for his failed ability to learn, I wanted to find an explanation for my own inability to teach him. In reality, the reason is likely somewhere in between, disproportionately affected by something in both of our lives that hopefully we can change together for the better.

The end result was that 25 percent of my children (one of four) ironically became the source of most of my perceived failure as a parent (this is an example of a skewed principle that I will describe later). In my son's skewed way of thinking, not only was he determined to simply forego higher education, but he was also willing to abandon our family's core values. He decided instead to mimic the local popular culture by more effortlessly finding a low-wage job requiring less education while regularly trying his random luck playing the state lottery, among a few other similarly nonproductive choices. As they say, "When in Rome, we should do as the Romans do."[92] In fact my son observed this general cultural disinterest in education everywhere around him (and copied it) and concluded that I was the oddball and not the norm. In a way, he was correct.[93]

---

[91] By analogy, I also infer here that my wife and I were unable to teach him a few vital core principles that could help him make appropriate decisions (like avoiding drugs). Like many families, we have not been as efficient as we would prefer in the outcomes we desire for our children, but we are still learning!

[92] This is credited to many, but the earliest quote I found was from an Italian pope (Clement XIV, a.k.a. Lorenzo Ganganelli): "When we are at Rome, we should do as the Romans do—cum Romano Romanus eris." It is also sometimes attributed to Saint Augustine from earlier in relation to following the customs of the local church rather than being dogmatic about subtle and unimportant differences. I like this because I am about to introduce you to another influential Italian, and I also agree we should seek the things that bind us together rather than divide us.

[93] West Virginia has the lowest level of advanced degrees of all fifty states, so he is right in concluding I am the oddball. Admittedly I am a bit odd (I would have to be in order to go to the extremes I have to

As he struggled within our local culture, I began to examine my own life imbalances, and while trying to find something useful (like a higher-paying career) for his own advancement, I discovered these repeating principles that were more widely applicable to all of us. These are all related concepts that I have discovered have similar reasons for their outcomes. But to understand them in the generic sense (i.e., basic human behavior), I needed to link these ideas with something we all could understand. So I therefore decided to begin this project by examining something all my children (and readers) could relate to—*money*, and more specifically, *income*. My kids, like Daisy, have all learned the relative utility of money in our culture, and I want to show it to them in the context of the modern American family (using mine as an example), which is why I chose *household income in the United States* as the basic model.

The Principle of Imbalance for American households (Income)

Presently, my *household income*[94] is consistently high and is in what is called the *fifth quintile* (a quintile is a grouping of 20 percent, of the population in this instance) when compared to the typical American household income group. Income clusters, more generally known as *quantiles*, are very imbalanced distributions within America (our household unit would be considered to be in the eightieth percentile or higher—meaning 80 percent or more families earn less than we do as a household). I suspected some form of imbalance, but the federal government actually has terms for examining this skewed income data since they have been aware of this

---

write this book). The norm in our state is a high school degree only. Bachelor's and master's degrees are disproportionately the lowest in West Virginia compared to the entire nation, according to American census data from 1990 to 2009. US Census Bureau, Table 233: Educational Attainment by State, *Statistical Abstract of the United States: 2012*. Only 17 percent of our population has achieved a college degree or higher, which is way below the average for the United States. So my son was right about the cultural norm not being higher educational attainment.

[94] This is also the case for my personal income. I felt household income was a better measure of the typical American family unit, so I use this measure mostly. I believe personal income is a component of household income, and it is easier to change household income (because only one of several people has to learn these principles) as opposed to an individual (you are the only one who can change). The two measures are very similar in their results of imbalance, though.

relationship for quite a long time.[95] I used to be even a bit embarrassed by this fact, especially living in such an impoverished state, but when I realized over twenty-five million households (roughly seventy-five million people) represent the top 20 percent, I felt a little better! So let me summarize what all this means.

I first examined the typical distribution of income for a given year to see how it is distributed or balanced within our country. I initially did this a few years ago using 2011 data and then later for 2012 (and several other years as well), and since I saw no appreciable change (it was always the same shape), I used the 2012 annual census data as shown. You can pick any year, and the relationship is roughly the same. This is an important point. *The imbalance is consistent.*

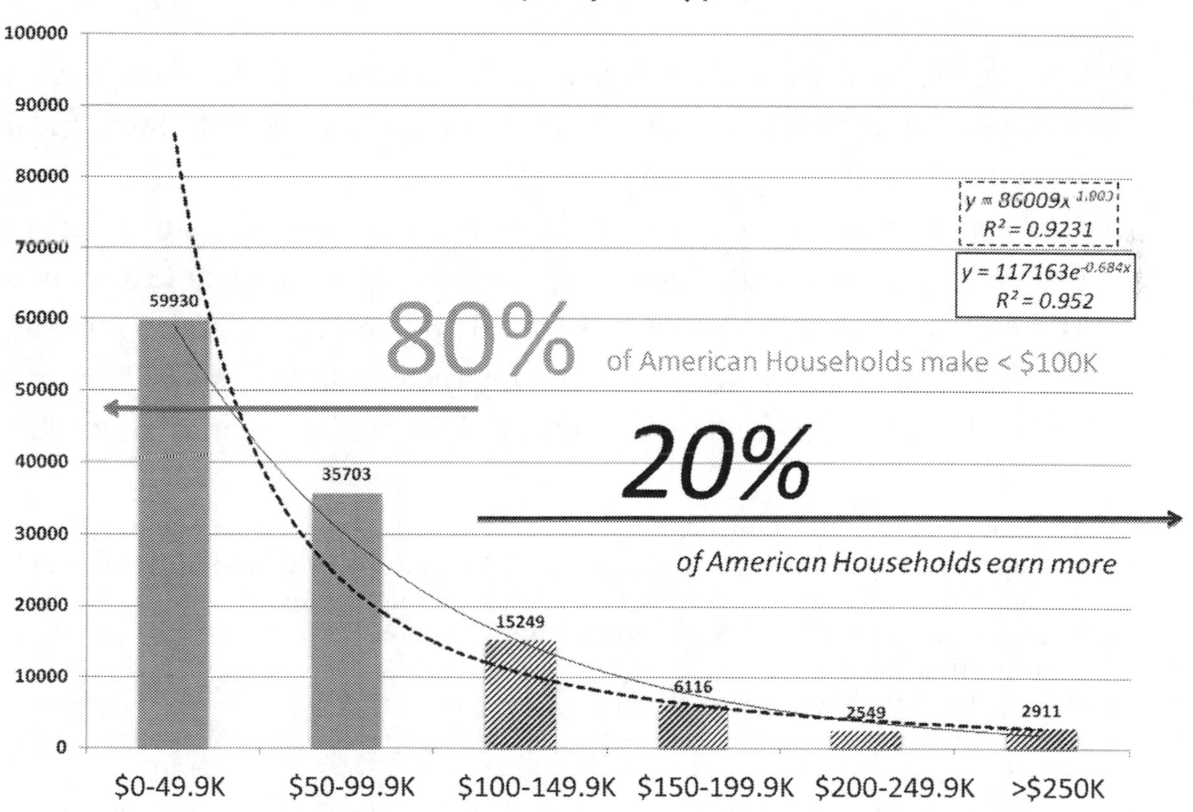

---

95 I know, for example, that Dr. Alba M. Edwards, one of the original census people for occupational data, joined the US Census Bureau in 1909 and created the hierarchal scheme of classifications for occupations (that we currently use) that nicely follows these rules of imbalance.

This graph shows the income distribution using US Census data[96] for all American households as increments of $50,000. The number of households earning within each of the levels is graphed on the left (y-axis in thousands) and the dollar amounts earned are on the bottom (x-axis). The curve[97] as graphed shows a distribution of the population within a range of $0 to greater than $250,000, where it is truncated for ease (we all like easy, including the US government). It is important to note that the money people earn is not spread out evenly. It is imbalanced, lopsided, or clustered within certain families. This is called a *skewed* distribution and is also known as a *Pareto distribution* of income (named for the person who first analytically described it),[98] and the numbers are relatively important (20 percent and 80 percent) as repeating relationships, as you will see.

A few comments are noteworthy from this graph:

1) The curve is very asymmetrical and imbalanced. It is in fact so abnormal and *atypical,* that it is in an entirely different category of distributions that are called *skewed.*
2) More people make little money. The most common group (first in rank or the most popular outcome) roughly earns an income between $0 and $50,000. They constitute roughly half of the entire population who earn money. This would be roughly the median household income since 50 percent would make more and 50 percent would make less

---

96 US Census Bureau, "Income, Expenditures, Poverty, & Wealth: Household Income," *Statistical Abstract of the United States: 2012*, last modified June 27, 2012, http://www.census.gov/compendia/statab/cats/income_expenditures_poverty_wealth/household_income.html. I soon discovered you could use any year for this data. I originally used 2011, and the cutoff at $100,000 was exactly 80 percent of the US population.

97 I plotted two best-fitting curves for a reason. In part this curve is probably exponential (the first part) and in the other part it is a power law (the tail part). The latter is what I find interesting, as I discuss more in part 2.

98 When I performed this analysis for 2011, it was roughly 80–20 using the groupings of $50,000, and for 2012 it was closer to 78–22 in the graphic, but you get the idea. The equivalent median income would now be closer to $52,000, and the top 20 percent would be closer to $105,000, but the *imbalance* is the idea, not the exact dollar figure. This is what greatly intrigued me. A few groupings (the first two in this case) account for most of the American population earnings for household income.

than this threshold amount. Likewise, the second group accounts for roughly 30 percent of the household income in America.[99] Two income groups (the first two) then account together cumulatively for 80 percent of the household population in America. Together they make very little.

3) A small percentage makes a lot. In fact, just how much is not shown since the US Census Bureau categorizes $250,000 or more as the highest category, so just how much farther to the right the curve continues is not evaluable here. In theory it goes on without a limit, although at some point there is only one household remaining at the top income level. This is important since this is where a relationship called a *power-law effect* is seen. The majority of the household income distribution is exponential in its mathematical skew,[100] but the tail section (roughly the top 1 percent) is guided by a different set of skewed math known as *power laws.* (Notice the power-law exponent in the above example is almost an inverse square, similar to gravity.) I will discuss the math soon.
4) The data was essentially the same regardless of the year I examined. In fact, if you were to examine this further (I do this later), you will find this same kind of imbalance goes back to the beginning of recorded descriptions of incomes in most countries and for most time periods, with little variation, as described by Pareto.
5) Making a lot of money does not seem like a popular outcome in America, although most people desire it. This is a good example of what I refer to later as a mismatch between what we desire and what we actually are able to achieve as a population. *We all want more, but we mostly get less.*

---

99 When I started this project, I used most recent 2011 data and $50,000 was the median roughly. In the few years it has taken me to get through the ideas I think are most relevant, it is now closer to $52,000 as the median, but the skew is still the same. It is always relatively skewed, and that is the point.

100 A. Christian Silva and Victor M. Yakovenko, "Temporal Evolution of the 'Thermal' and 'Super Thermal' Income Classes in the USA during 1983–2001," *Europhysics Letters* 69, no. 2 (2005): 304–10. Exponential to me implies simpler supply-and-demand growth principles whereas the mechanics of power laws are a different sort altogether.

6) If you wish to assign numbers to it, roughly 80 percent of the household population earns just a little money, and 20 percent (the top quintile) of the population earns a lot. This imbalance is what Pareto first described in his country. He similarly showed this same *principle of imbalance* for wealth accumulation, which is even more skewed than simpler household income. [101]

The income data I examined and used for this graphic image was from the last decade in America, but *it rarely ever changes on the whole*, as I saw for the years I examined. While it may change in the degree slightly, the principle remains the same—*income is always naturally skewed in this way.*

My next logical question therefore was, "Why is it skewed?" or more relevantly, "Why is income skewed here in America and why now in the twenty-first century?"

Income Is Always Imbalanced—Here, There, and Everywhere!

The curve for income distribution for modern-day America fits what is known as a *Pareto distribution*, which is a type of mathematically skewed outcome that relates to money. As it turns out, this imbalanced distribution of income is neither modern nor American.

Pareto, an Italian engineer, first described this idea over one hundred years ago on another part of the planet and within a relatively different system of social and economic governance.[102] This was intriguing to me. I assumed income imbalance was mostly an American issue and therefore related mostly to capitalism.[103] I also assumed it was a modern phenomenon

[101] Wealth depends on income and then excess income beyond expenses (i.e., savings), which then grows by another power-law phenomenon called compounding interest, which means your money acts as its own catalyst to make more money (without you doing the work).

[102] The point here is that the geography, the politics, and the leaders don't really matter for the overall concept of the imbalance—it is simply human nature. What matters is the degree of this imbalance, as I will discuss later.

[103] Never assume. You know what they say. This is a good point where assuming is wrong. Why should income imbalance be American in concept—we are not so special! It is more rationally a product of human behavior, and while we may be better at separating income as a society, I can assure you it is not an American concept.

and not one that repeated itself throughout human history. I was wrong on all accounts.

After spending a lot of time trying to make heads or tails[104] out of what all of this means mathematically, I more simply (more efficiently) went back to this engineer-turned-economist's original description and the similar imbalance he noted in Italy from the 1890s. This provided a treasure of knowledge, which I will try to efficiently summarize for you in just a few pages.

Pareto was an analytical man who believed human behavior could be approached much like other physical sciences; he believed that human behavior could be analyzed with descriptive formulas and rules similar to those seen with astronomy and other physical and mechanical universal forces.

What Pareto demonstrated was that people more commonly make little income and fewer people infrequently make a lot of income. Pareto described this imbalance using the analogy of a *social pyramid* since it became increasingly harder to reach the top income levels. He consistently found there to be a repeating nonlinear shape (now known as a power-law curve) for income and wealth, later suggesting it could be applied to any human ability.

To verify his hypothesis further, Pareto critically examined this unusual repeating nonlinear relationship between income and people by tabulating various income distributions for several other countries (e.g., Peru, France, Germany, Ireland, and Great Britain) and for many other time periods (1400s, 1600s, 1890s), and he discovered the same predictable imbalances. His tabular data from Great Britain from 1893 is graphed here as an example to show the same skewed relationship seen elsewhere.[105]

---

[104] Ninety-nine percent of the distribution is exponential in nature (the heads part). The top 1 percent is where the action is, and there the math changes to power-law relationships (the tail part), which are much steeper and much more skewed. For example, the range of income for the top 1 percent may be anywhere from $400,000 to $40 million, which is huge, whereas the other groups have more finite ranges. Things that vary by such huge magnitudes almost always are related by power laws, which I discuss in more detail later.

[105] He examined tax records of Basel, Switzerland, from 1454 and from Augsburg, Germany, in 1471, 1498, and 1512; he looked at contemporary rental income from Paris and personal income from

Could this repeating imbalance from a century ago in another nation in fact just be a coincidence?[106]

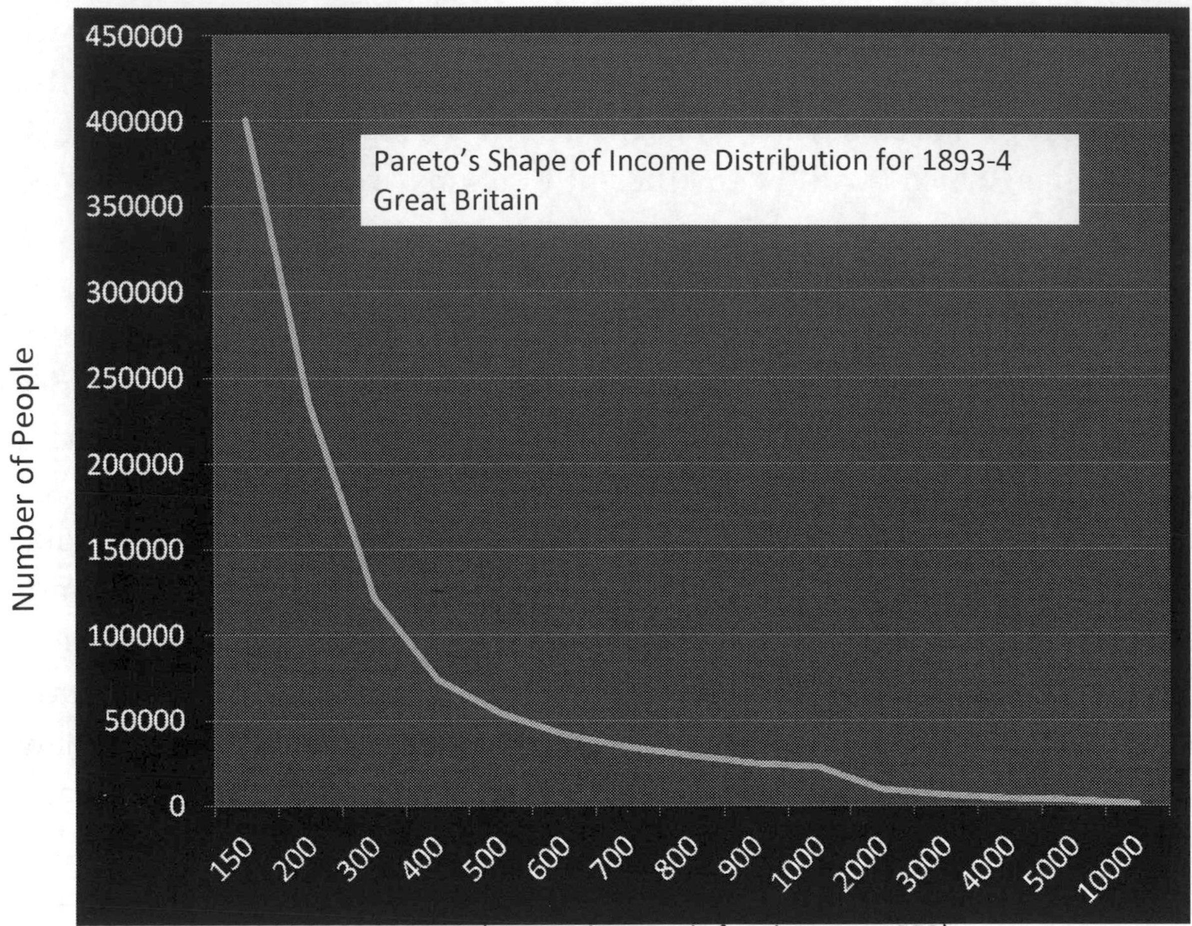

The inequality in distribution of income in Great Britain during Pareto's time period is clearly imbalanced, and it follows a similar skew as today's income imbalance within America. In this case roughly 80 percent of the British population earned an income of £400 a year or less, while only 20 percent made £500 or more (and up to £10,000). These numbers are

Britain (as represented in my graph), Prussia, Saxony, Ireland, Italy, and Peru. The results were the same.

106 Pareto (who appropriately did not believe in random coincidence) was more exact and believed in describing these ideas as methodically as possible: "Science tries to bring theory as close to the facts as possible, knowing very well that absolute coincidence cannot be attained. If, in view of that impossibility, anyone refuses to be satisfied with approximate exactness, he had better emigrate from this concrete world, for it has nothing better to offer." Pareto, *The Mind and Society*, 1:64.

consistently imbalanced in a similar way in most time periods[107] and everywhere in the world.

The remarkable similarity of income imbalance for so many different countries (socially and politically) and for repeating time periods led Pareto to conclude over a century ago that income and wealth are imbalanced predictably "through any human society, in any age, or country."[108]

This became in fact a social law for income imbalance and is probably something that every economist in the world knows. It is now referred to as *Pareto's principle* or *the principle of imbalance*. Had you ever heard of it before this book? *Me neither!* But then I am neither an economist nor a sociologist—I am just a casual but interested observer and a modern American thinker.

While I could not initially explain the mechanics of the skewed mathematical principles of observed income spread, what soon amazed me was the repetition throughout various geographies and time periods. The ubiquitous nature of this skew is what first caught my analytical eye—not so much the weirdness of the skew (though that caught my attention too, but that is complex math). In other words, *this is not a modern or local phenomenon.* It repeats everywhere and all the time—and predictably so. It occurs in non-American cultures that have differing governments and thus different policies—and yes, even in countries we think of as having different societal (e.g., religious, economic, and political) values.[109]

---

[107] In this case, 37 percent of the population made only £150 or less, and the median income was somewhere between £150 and £200. A few people made a lot, and most people made only a little. The table I used was from income in Great Britain 1893–1894 tabular data from Vilfredo Pareto, which I graphed from Vilfredo Pareto, Schedule D Table, *Cours d'économie politique*, ed. F. Pichon Paris, tome II (1897), 305.

[108] V. Pareto, "The New Theories of Economics" *Journal of Political Economy* 5, no. 4 (September 1897): 501. This implies the culture is less influential than human nature or our abilities since Pareto showed the imbalance in diverse societies like ancient Peru, Cherokee in America, and Prussia. Later he suggests since these laws are true everywhere, changing the distribution of the income does not help; instead our efforts should be on increasing the mean income.

[109] I don't know about you, but I don't consider myself remotely Italian. Yet it would seem we are not much different. In fact, this is something I have concluded many times over in my life through my travels and that I hope to model to my kids—not many people are in fact different anywhere in the world. We mostly all want the same basic things. And while there is always an exception to the rule,

Therefore income imbalance is more likely related to the nature of human development, perhaps amplified through the efficiencies of industrialization and globalization, than in our unique policies or our so-called governing elite (our nurture). That is not to say the degree cannot change (this is in fact part of the modern-day discussion in our country and will also be seen around the world).[110] But what is relevant to us is that this unequal balance is truly a universal concept.

In addition to quantitatively setting out to describe economics, Pareto was a truly gifted man with an affinity for knowledge.[111] He spoke seven languages and familiarized himself with all leaders in academics including philosophy, religion, and human history. He posited that everything could be analyzed through the scientific method, including human behavior. With all of this background, he began to methodically describe what became the basis for a new science of human behavior, which later became the foundations of sociology. Today Vilfredo Pareto would be considered a true Renaissance man or a polymath, similar to American founding fathers Benjamin Franklin or my favorite, Thomas Jefferson.[112]

Although originally an engineer (he would therefore represent the modern version of a *tinker* [e.g., what I subjectively define as an engineer, scientist, inventor, architect, etc.] for the first twenty years of his working life), Pareto later acquired a fascination for socioeconomics, and he would spend the remainder of his life developing his ideas into scientific theories, setting the future stage for thinkers in these areas. With his rigorous mind for analytical math and the scientific method, Pareto meticulously recorded

---

we must try to remember that no one is better than anyone else in the world, although some may be better off.

110 According to Pareto, the exponent or degree of the distribution was usually between one and two, as it is in the graph I showed above. The degree of the skew is a function of what each society choses to value (more or less) in my humble opinion.

111 His bachelor's was in mathematical sciences in 1867, and he received a PhD in engineering in 1870, with a dissertation on "The Fundamental Principles of Equilibrium in Solid Bodies" (Turin Polytechnic University for both degrees). So technically he was a tinker and a doctor.

112 Both American men are coincidentally found on our disproportionately distributed denominations of money (the two-dollar bill and the hundred-dollar bill), and needless to say both men were profound contributors to American culture through their careers and accomplishments. Jefferson was even the commander in *chief* as well as being a lawyer, merchant, and tinker (he was an inventor).

data for income in various countries and at various time periods and showed similar occurrences of this imbalance of income and wealth.

The principle of imbalance consistently shows a predictably imbalanced relationship between the various inputs and outputs—in this case, people and money. One is a *power function* of the other, similar to gravity with its repeating predictable power of 2, called an inverse square. While Pareto described the relationship using proper complex mathematics (he preferred to use related ideas of logarithms[113]), what he found most important was that it repeated cyclically in both space and throughout time; in other words, space and time become irrelevant to it, much like what is observed in many other physical principles of the universe.

What I like to remember from his work is not the complex math but the simple visual imagery of the skewed curve I showed you previously. This weird, lopsided curve is something that predictably defines human behavior in more ways than just for income. Pareto suggested this principle of imbalance was a measure of any relative difference in human ability (or as I prefer to think, universal efficiency).[114] As I tell my children, "Life is mostly nonlinear and usually very skewed. Try to learn what side of the curve to be on early if you are able."

In his many writings, Pareto attempted to explain the reasons for these consistently skewed outcomes, citing differences in our human capabilities, as well as the differences in various governments and their ensuing policies (since they are just groups of the same differing people).[115] One of the explanations Pareto offered was the idea of hierarchal thinking (i.e., a system of rank) in the way we perceive everything relative to us, and he used this idea to describe what he referred to as the *social pyramid*[116] of human nature.

---

[113] This is often plotted on log-log axes, which would result in a linear graph, one hallmark of these true power laws. One of the first power laws in fact related to an astute observer noticing certain pages from log tables (that is what we had before calculators and computers) were more commonly worn by use.

[114] To Pareto, "They [natural laws] all arise in a desire to give a semblance of absoluteness and objectivity to what is relative and subjective." Pareto, *The Mind and Society*, 1:265.

[115] He also believed people are motivated less by logic and reason and more by sentiment and emotion.

[116] Although it is sometimes referred to as a social pyramid, he makes the analogy that income distribution is more like a social arrow—very fat on the bottom where the mass of men live (poor)

Pareto used the simple analogy of a *scoring system* to show how society invisibly rates individuals within each brand of human activity—a type of internal ranking system from 0 to 10 that creates a hierarchal classification system breeding what he called *elite*[117] and *non-elite* classes. He also compared it to the way we grade our children in school, which is in truth one of the earliest versions of social stratification that our society still teaches through academic achievement.

In essence, Pareto was stating that what we tend to do in society is classify and rank what we all value and do as humans from early childhood on.[118] How we then reward this hierarchal achievement is naturally and disproportionately based on this systematic rank of classification. Rank distributions in general (and by no coincidence) usually seem to follow similar power-law distributions, which are predictably imbalanced according to the same skewed math and how we attach value disproportionately to the various subsets.

Pareto offers a few popular examples showing social stratifications with "lawyers, poets, chess champions, and thieves" and even "women seducing powerful men."[119] These people were essentially ranked in society based on their unique capabilities, with the top achievers forming the elite classes within each category. According to Pareto, "The highest type of lawyer, for instance would be given a 10. The man who does not get a client will be given a 1—reserving zero for the man who is an out-and-out idiot." So in

---

and very thin at the top where the wealthy and elite sit. This was a repeated social law, not by chance, and was "in the nature of men." Pareto, *Cours d'économie politique*, 313–18.

[117] Per Pareto, "In a broad sense I mean by the elite in a society people who possess in marked degree qualities of intelligence, character, skill, capacity, of whatever kind ... On the other hand I entirely avoid any sort of judgment on the merits and utility of such classes." Pareto, *The Mind and Society*, 3:1421.

[118] Pareto's *The Mind and Society* is in fact one of the first textbooks in sociology to explain the relationship between the *human mind* and *societal* outcomes using the scientific method, thus his choice in a title. It amazingly parallels our world today, despite a different political system, a different geography, and a different time period.

[119] "To a clever rascal who knows how to fool people and still keep clear of the penitentiary, we shall give 8, 9, or 10, according to the number of geese he has plucked and the amount of money he has been able to get out of them. To the sneak-thief who snatches a piece of silver from a restaurant table and runs away into the arms of a policeman, we shall give 1." Ibid., 3:1422.

simple terms the scale (for lawyers' abilities) ranged from idiot—0 to the elite—10.[120]

Similarly he noted, "To the woman 'in politics' ... who has managed to infatuate a man of power and play a part in the man's career, we shall give some higher number, such as 8 or 9; to the strumpet who merely satisfies the senses of such a man and exerts no influence on public affairs, we shall give zero." [121]

And in reference to wealth, he said, "To the man who has made his millions—honestly or dishonestly as the case may be—we will give 10. To the man that has earned his thousands we will give 6; to such as just manage to keep out of the poor-house, 1, keeping zero for those who get in ... [To] those that have the highest indices in their branch of activity, ... we give the name of elite ... And so on for all of the branches of human activity."[122]

Pareto believed this natural comparative system of hierarchy applied to politics, careers, education, sports, music, religion, and income—to anything where human social grouping was involved—indeed even to the system of scientifically describing the model itself.[123] Furthermore, within governing clusters (where people are able to therefore exert power and influence) these elite members would collectively constitute a *governing elite* if they were able to influence policies (there would also be a *non-governing elite*,[124] as well as a lower stratum called the *non-elite*, which would be most of society). Any long-lasting governing elite was referred to as an *aristocracy*.[125] Many would argue that the elite's ability

---

120 Ibid. If you were that well trained or well educated and could not get a client (and hence have no work), it would imply you were an idiot relative to everyone else. While this may seem rudely inappropriate in today's world, the point he was making was all about relativity. The more-modern jargon might be the *less fortunate* rather than *idiot* (and likewise, *more fortunate* rather than *elite*), but for my lawyer friends I do admit it has a relatively comical connotation.

121 Ibid.

122 Ibid.

123 What I infer here is that the very nature of classification tends to place things of similarity with others. Therefore it is a human concept to classify, and there is therefore an inherent method that causes these outcomes to look this way based on how we think.

124 Ibid., 3:1423. Pareto (who was born French) borrows this idea from Kolabinska, "La circulation des élites en France," page 5, which he references.

125 Pareto's father was an aristocrat that fled Italy to Paris, which is how Pareto was born French. Later when political tensions eased, the family returned to Italy, but I am certain these shifting

to shape governing policies is an unfair advantage, but in most societies where this was propagated unfairly and for long periods, this system was destined for failure according to Pareto. Furthermore, according to Pareto's analysis, within all societies, these elite (and non-elite) groups circulate in such a way that a continuous state of rapid equilibrium[126] is formed.

Socially what he implied was a dynamic and cyclical reproducible model of human behavior categorized (and ranked) by some systemic internal hierarchal classification—much of this from his simple observed model of skewed income distribution![127] Modern social economists debate his equilibrium model of cyclical social circulation, but what he described is simply the concept of social-class mobility, which I believe is a repeating human principle that he nailed with accuracy (I would also suggest a similar circulation between American governing elite Republicans and Democrats).[128] Upward social mobility is the essence of what many Americans in fact hope to achieve with the so-called American dream—the ability to move from rags to riches. After all, what is the point of living

---

ideas helped shape Pareto's views on life. His father was also an engineer (another tinker), and this likely disproportionately factored into Pareto's choice of initial careers as well.

[126] Ibid., 3:1430. Pareto famously said, "History is a graveyard of aristocracies," implying that the governing elite repeat the same mistakes over and over. Low-quality elements accumulate in the elite, while high-quality elements in non-elite are prevented from rising. Elements not needed to maintain power accumulate in the elite, and these traits compromise the circulation of the elite, leading to revolution, change, and eventually a new victorious revolutionary leader that becomes the new elite (who then starts the process all over). It is therefore a never-ending vicious cycle of human behavior but also one that in theory results in improved quality and efficiency over time. Pareto cited the Athenians, Romans, Barbarians and Frankish conquerors, English nobility and descendants from William the Conqueror, and German aristocracy (Ludwig?) as examples.

[127] I am certain Pareto developed some of this mathematical model from related *indifference curves*, from another mathematician from an earlier time period, F. Y. Edgeworth. Edgeworth describes his own *principle of greatest pleasure* (for ethics and economics) as analogous to the *principle of maximum energy*. Edgeworth describes the concept of efficiency that I am after as well. He also refers directly to the idea of social mechanics being similar to celestial mechanics, which is what perhaps inspired Pareto among other ideas. He also describes the ideas of supply-and-demand curves (they were called contract curves) and references Leon Walras, who preceded Pareto as professor of these same economic equilibrium ideas at Lausanne University. F. Y. Edgeworth, *Mathematical Psychics* (1881), 12–13.

[128] One idea he is criticized for is his opposition to democracy, which he considered a failure.

if a person cannot aspire to reach the pinnacle of his or her own God-given talents and abilities?[129]

Before trying to blame the principle of imbalance on the Italians and their dominant Roman influence on our culture, which was my first reaction, I should point out that this data has been examined for many other non-European cultures and for other time periods as well. I am no expert historian, and thus I cannot attribute all behavior (i.e., inputs) to the Greco-Roman influences on the many world cultures of the modern era, though they undoubtedly factor in somehow disproportionately based on many events and how we incorporate what we have learned (i.e., outcomes) within the context of the evolution of ancient human history.[130]

Other critical thinkers became interested in this consistent principle of imbalance, including German American educator George Zipf, who corroborated this data with his own data within early America. Therefore, regardless of where you live,[131] when you live, or what governing culture you live under, this same principle of imbalance naturally applies to you! And while the amount of (money) imbalance may vary in degree some between countries,[132] it represents simple skewed human nature and our desire to classify and rank it—with income as the easiest measurable result.

---

[129] The implication here is more than money. The implied growth is in spiritual development or in our abilities to give to society and improve it somehow.

[130] For example, after initially persecuting Judeo-Christians, Romans adopted Christianity as a more efficient teaching tool of morality than previous pagan beliefs. Later, the Roman governing elite would even rewrite history and renumber our calendars (with base 10 Roman counting, as well as using Roman alphabets) in an effort to record history in relation to the life and death and resurrection of Jesus Christ. Whether this was accurately done or not, their culture was efficient at assimilating others and adopting ideas that they then propagated using these same principles. I suggest Stephen Jay Gould's book *Questioning the Millennium: A Rationalist's Guide to a Precisely Arbitrary Countdown* (New York: Harmony Books, 1997) as a reference to some of these ideas of the Romans' great but subtle influences on our everyday concepts of time. Of course much of our language is Latin based as well (as is our base 10 counting system), which has a lot to do with these origins from our Italian friends.

[131] I could argue that certain cultures have learned to magnify the imbalance more than others (whether that is good or bad), which is propagated as part of the culture as it is here in America.

[132] Some countries have a little less or a little more inequalities, but they are always the result of human nature. Scandinavian countries as a rule seem to enjoy a happier balance of this inequality than most other countries, suggesting a model worth learning from in efficiencies (if this were your desired outcome).

I don't want you to become overly concerned about the details of the math here—just get the concept that *money is something that is very imbalanced in our skewed world.* Just *how* imbalanced is a different matter. Society refers to this idea as *income inequality* (or income disparity), and it turns out to be just one example of many other similar observations that exist in our *skewed* world culture. *America (and similarly the world) is naturally skewed, and if you are able to learn this now, you won't feel as if you got screwed later.*

In the next chapter, I wish to show you another example of this same principle of imbalance as it applies to the many occupational choices within America through the related laws of supply and demand.

"Doctor, Lawyer, Merchant, Chief."

The Principle of Imbalance and Your Use of Time

As parents for almost twenty years, my wife and I have tried to help our children find true passion in their unique interests and to help them recognize their individual strengths by further exploring them while they are young. The hope of course is that they can learn to combine this with career choices and live happier lives. For each child, this differs, and part of the fun and challenge for us in having four children is that we were able to explore a lot!

Time, whether you realize it, is very imbalanced for us all, and when and if you have children, it becomes even more so! Likewise, as you get older, you will appreciate this temporal inequality[133] even more. Just like a rhyme with a reason, there is a similar imbalance with our time and a season for everything we choose to do.

If you were to reflect on your youth with the power of 20/20 hindsight and analyze how you allocated your time, you would see that school and related activities (e.g., homework) formed an unequal, large part of your time as a youth. In a similar way, as we enter into adulthood, we find our time disproportionately allocated to our occupation and work-related activities (getting ready for work, unwinding after work, planning work at home, etc.). To illustrate this, I've laid out my own typical day in the below graph to show where my time is most disproportionately allocated.

[133] I would point out that our concept of the age of the universe is also skewed since we think of time in linear, equally spaced intervals relative to other concepts like forces in the universe. There is no reason not to think of time as nonlinear beginning with the idea of the original frame of reference (creation or the big bang), and therefore time would be defined differently as it progressed along a curve rather than a time line. This is also relevant to people who read the Bible and who try to make creation events fit linear-time concepts. It may be that 80 percent of the universe was efficiently created in 20 percent of the time (as an input) and the rest is trivial.

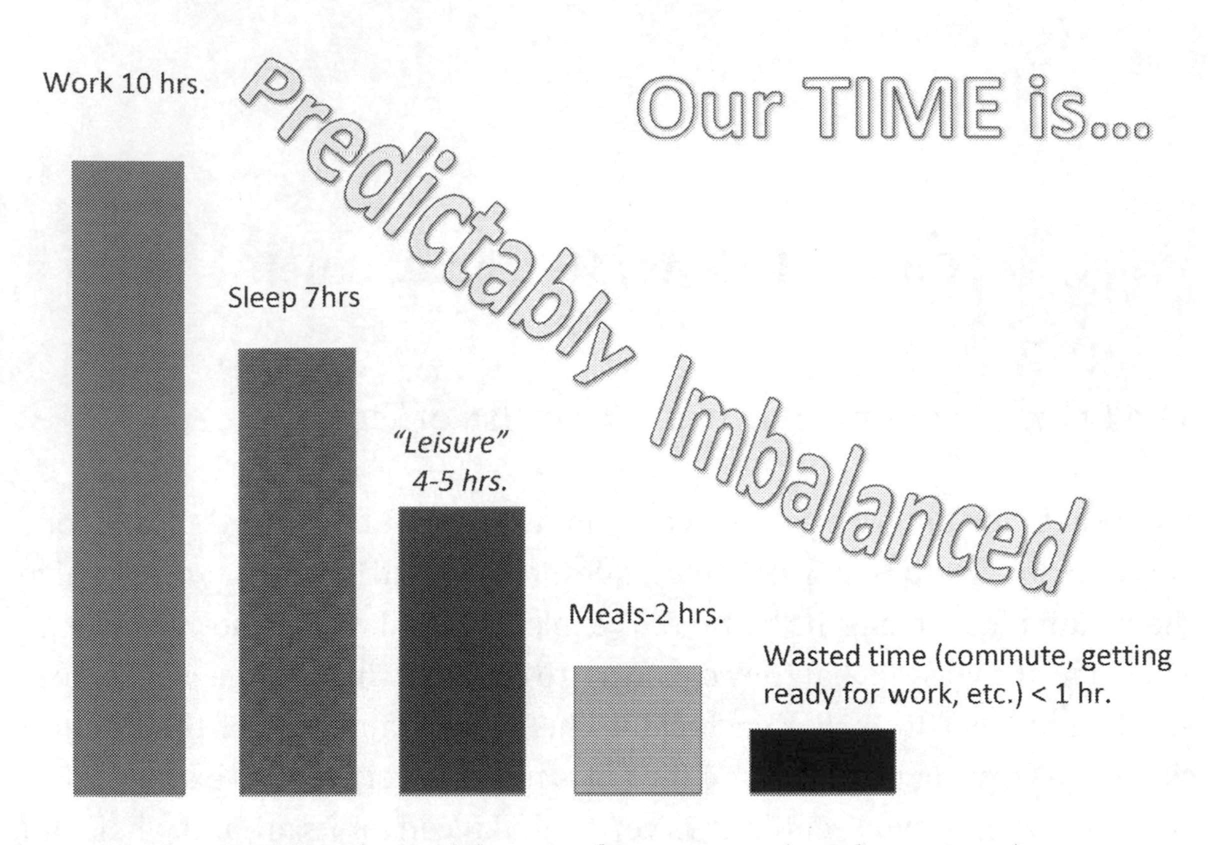

How TIME IS Distributed for me for a typical 24 hr. period

As you can observe from this simple graphic, most of my time is spent *working* as an adult—and I used a conservative estimate of my time. For my children, I would substitute "school" for "work" in this graph. Sleep is required by all of us, so once I examine the waking hours, I find that I spend roughly 60 percent of awake time at work and only 20–30 percent of time is left over to enjoy leisure and other fun (the other time is "wasted time" or consumed with eating and grooming). You might also be able to infer that if you have children, the time gets allocated even more disproportionately.

Since *time working* disproportionately accounts for our awake time (assuming we are employed), it seems logical and efficient that this input factor should be well planned, well researched, and well chosen to maximize happiness and well-being.[134] This is an area where most people naturally

[134] Another way to think of this is to say that one variable (e.g., leisure time with family and friends) accounts for most of my fun and the majority of my happiness. Let's say there are five variables. So

struggle for lots of reasons. If you can learn to spend most of your time doing what you love as an occupation (and it is your passion), you will feel blessed as I do since you will be happy most if not all of your day.

Because work consumes so much of our time, it was one of the first variables I decided to examine for my children. I decided to analyze it for the simple measure of income, as it is more objective data, although I could have also chosen to use something more subjective like happiness or job satisfaction. The results would be the same.

Income Imbalance by Occupation

Initially, I started my income-by-occupation analysis at the top (after all, I think in terms of rank as most humans do). I first chose to examine income data for the *top twenty occupations* as individual entities by popular outcomes (in this case using the tool *money*). I used the US Bureau of Labor's annual report based on the occupational database reported from 2011 (2012 was the year I began this project), which examined the eight-hundred-plus individually tabulated occupations within America. I based my initial analysis on *median annual income*,[135] but you could also use hourly wages

then one variable of five, or 20 percent of my daily inputs, accounts for 80 percent of my happiness. The variable could be food or your career; the relationship remains the same—one variable of the five (20 percent) would account for most of the happiness. The point is not the variable but realizing the relationship to what you desire and the various inputs. For example, people who love their jobs will spend even more time at work and become workaholics. Similarly people who hate their jobs or the other variables may substitute eating and become foodaholics, or they may become depressed, sleep more, or drink more. The relational ideas are the same. Similarly you need to balance exercise, church, friends, time with kids, spouse, etc., with consistent efforts in the rapidly diminishing time you have left over each day.

[135] Please note this is individual occupational income, which is a part of why household income is also imbalanced. Household takes into account your occupation as well as how many people live together and thus is more than just how much you make as an individual. This fact is very important since the skew does not seem as bad for occupational income—in fact, occupational income seems simply exponential and not a power law, which to me suggests purely supply-and-demand rules at work based on population and little else. Household income becomes more skewed because of human choices—like being married (and having two income earners) or living with someone else who also earns income that then gets counted into the household income. This is in fact one of the main reasons we have seen household income go up for the top earners in America. The top-earner families are more likely to have multiple income earners than the lower-income quintiles.

or some other measure if you so desired (e.g., percentiles, average incomes, deciles, quantiles, etc.). The measure does not matter, since it is more about the relative relationships here anyway.

The following are the top occupations ranked individually in order from one through twenty, from recent US (2011) Census Bureau of Labor data:

1. Anesthesiologist ($234,950)
2. Surgeon ($231,550)
3. Obstetrician/gynecologist ($218,610)
4. Oral and maxillofacial surgeon ($217,380)
5. Orthodontists ($204,670)
6. Internists, general ($189,210)
7. Physicians and surgeons, all others ($184,750)
8. Family and general practitioners ($177,330)
9. Chief executives ($176,550)
10. Psychiatrists ($174,170)
11. Pediatricians ($168,650)
12. Dentists, all other specialists ($168,000)
13. Dentists, general ($161,750)
14. Petroleum engineers ($138,980)
15. Podiatrists ($133,870)
16. Prosthodontists ($130,820)
17. Lawyers ($130,490)
18. Architectural and engineering managers ($129,350)
19. Natural science managers ($128,230)
20. Marketing managers ($126,190)

A noteworthy prevalence of certain occupations here surprised even me at first glance. The published data I used was initially in 2012 and from the 2011 census, but the trends are the same year to year, so don't focus on the year or even the dollar amounts—I just want you to see the relative rank. I looked back over the last twenty years, and the relative positions changed very little by year, and the occupations remained the same by preference. The same career choices seem to be consistently efficient performers! As a reference for comparison, during these twenty years, the average annual salary for all jobs (after adjusted for inflation into the same 2011 dollars) was $42,194.[136]

At first glance, I simply thought this was mildy interesting and probably based on supply-and-demand economic efficiencies, but the results seem to repeat in strange ways that begged some further analytical explanation. Why would the same occupations always do well and why in these proportions?

Examining the top twenty occupations, I next noticed there was a disproportionate overlap with one profession—the *doctor.* There was in fact a noteworthy *imbalance of professions* within the top twenty occupations consistently each year. Of the top twenty jobs, there were four to five consistent categories, and of these broader groupings, just one category (25 percent of the groups) accounted for most (70 percent) of the top-paying jobs.[137]

Over the last few decades the top career outcomes (using income as the endpoint) fell into the following four categories:[138]

1. Medical and dental (professional) [*doctor*]
2. Business and technology management (and CEOs) [*merchant* and *chief*]
3. Law (and related) [*lawyer*]
4. Engineering/science [*tinker*]

Doctors of medicine and dentistry repeatedly account for fourteen of the top twenty top earners (70 percent) and fourteen of the top sixteen spots from the 2011 data; business and technology management consistently account for four of the twenty top earners (20 percent); lawyers take one spot (5 percent); and engineers make up one to two (5 percent) of the top-earners spots.

---

[136] US Census Bureau, Table P-43: Workers (Both Sexes Combined—All) Median and Mean Earnings: 1974 to 2011, *Historical Income Tables.*

[137] This relationship is an approximate example of the 80–20 principle that I discuss further in later chapters that essentially says a vital few inputs are responsible for most of the outcome results, in this case by ranking of incomes and careers. In this case it would be a 25-70 relationship.

[138] These would seem to fall nicely into the categories of doctors (in dentistry or prosthodontics or medicine), lawyers, merchants (business and management), and chiefs (CEOs) without too much effort.

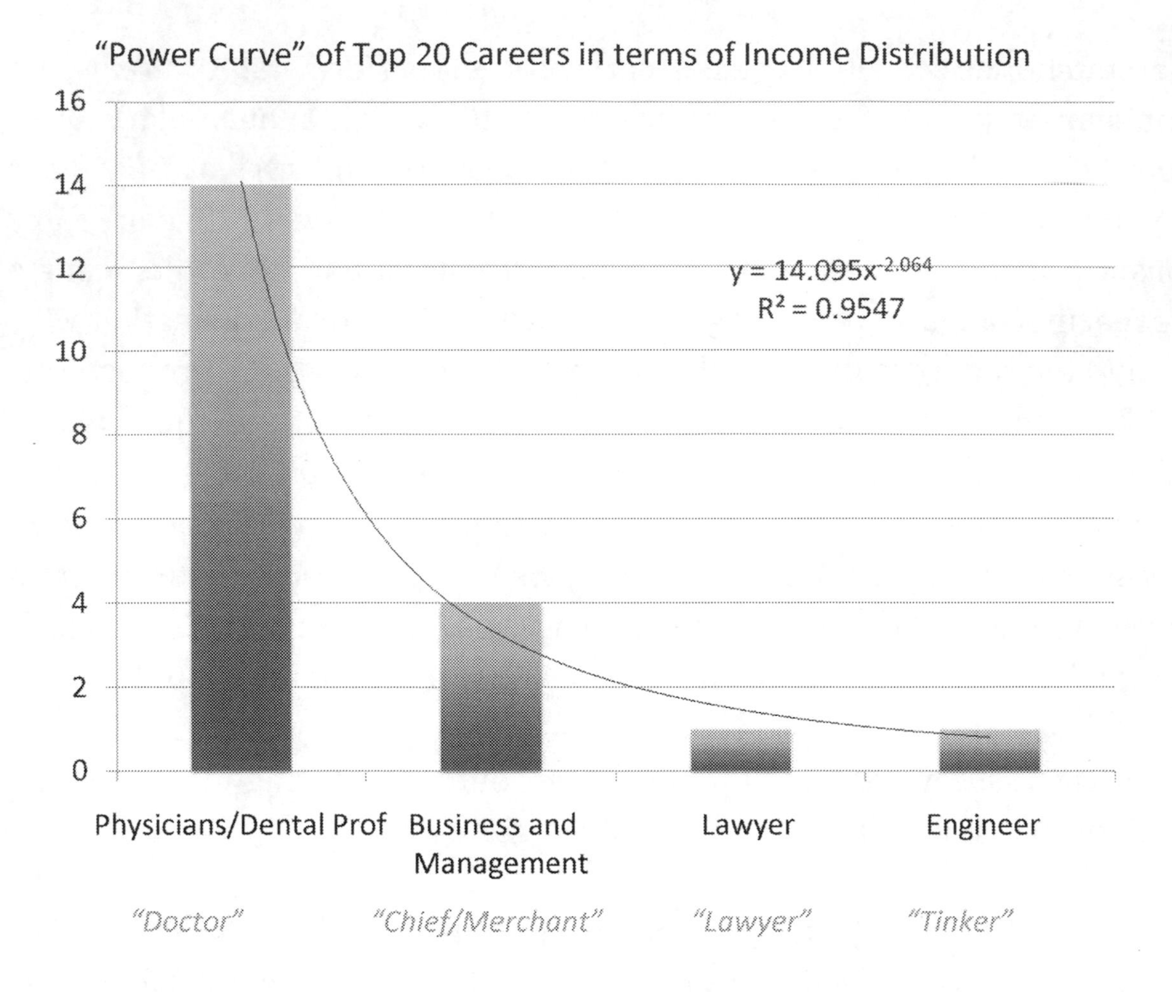

The top twenty occupations as listed by the US Census Bureau in the previous table (based on the same 2011 data) are shown here as a histogram. When plotted by rank in order, the data seems to follow a distribution that is a familiar imbalance, the Pareto distribution.[139] The surprising majority are medical professional occupations (including general medicine, various specialties, and oral and dental health-care practitioners). The list is by no means complete, but the concept of proportionality or imbalance is what I am after here. Notice it is roughly an inverse square relationship, similar to the unchanging law of gravity.[140]

[139] Rank distributions are in fact often disproportional and skewed like this. How we rank outcomes is in fact part of the very issue. We prefer these careers as a society and reward them handsomely and disproportionately. Here I used Microsoft Excel to generate a best-fitting line (curve), which is a power-law curve.

[140] One might hypothesize people so inclined in these abilities (to borrow Pareto's term) are *naturally attracted* to these professions.

Stated more generally, a few inputs (occupations) here account for the majority of the highly rewarded outcomes (money). And furthermore within these few categories (four are shown here), roughly 25 percent of the inputs (one group) disproportionately accounted for most of the high-income results (I could likewise show something similar on the other end of the spectrum with low-paying jobs).

A few occupations consistently account for the majority of high-paying jobs for Americans. As you can see, the nursery rhyme was not too far off from the truth, even a few hundred years ago when it was first created! Things don't seem to really change (with the exception of Daisy now also being the primary breadwinner as well as potentially marrying someone with earning power).[141]

One parenthetic note to mention here is that these statistics don't account for independent business owners and self-employed career earners. These occupations are not included in the statistics[142] by the Census Bureau for reasons that are unclear to me.

The main point is that these same professions have been around a long time and are not likely to change a lot in the near future (perhaps with the only exception being a newcomer connected to the computer industry and related spinoffs with Internet and software). Society consistently values these professions disproportionately, either by laws of supply and demand, or for other rhyme and reason, and there is not a lot of supply[143] for the demand, as we will see (due to educational-attainment shortfalls). If you were to choose any one of these careers today, I am confident you would still look back in twenty years and feel like a top performer relative to all other careers. Daisy clearly knew her stuff!

---

[141] I don't have the time or desire to go into the disparity between male and female. I know it exists, but I really do think this issue will resolve sooner than the other ideas I am presenting here. I wish to emphasize to my Daisys that they should learn to take care of themselves.

[142] Also note that some of the salary figures may not reflect true earnings. For example, chief executives may have other compensation packages that are not reflected here (e.g., stock options, production bonuses, or retirement parachute packages), so the data is not perfect. Other bonuses and production earnings may not be factored in here as well. But the trend is quite obvious.

[143] I do go through the additional effort to show the supply of labor in the appendix of this book, but it has to do with more than just supply and demand. Educational time and expense commitments and minimal levels of achievement are also intimately related.

So next I asked myself, "If these are the occupations that consistently pay disproportionately, what do people in America actually choose? Or more relevantly, why don't more people choose one of these careers?"

"Popular" Occupations

Next I decided to examine the distribution of the population within the various categories of occupations in the United States. The following are the occupations we choose by popular outcome within America, despite our expectations to achieve the outcomes of the preferred and highly ranked careers I showed you previously:

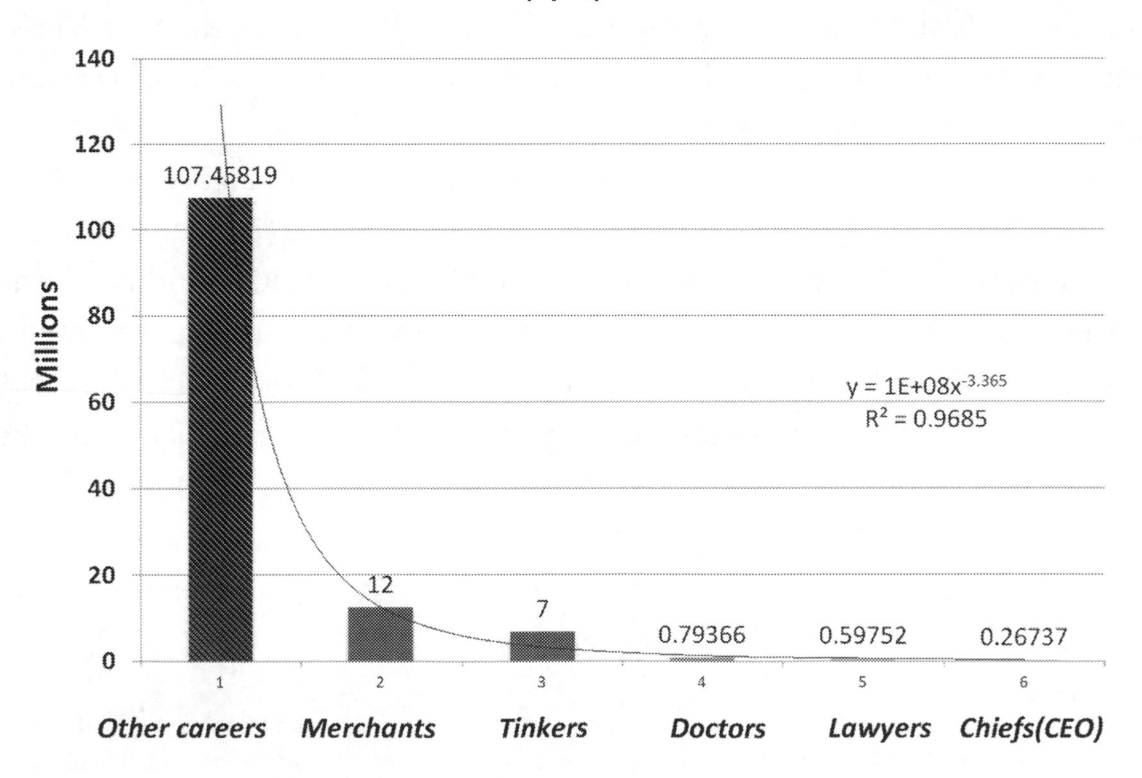

This graphic shows the number of people (in millions) in all the various occupational categories as defined by the US Census Department of Labor. Most Americans *do not choose* the twenty top-earning occupations, either from desire or through ability. As shown in percentages, 84 percent of

Americans choose other careers besides doctor, lawyer, merchant, or chief, while 10 percent choose to be managers (i.e., merchants and related business/ finance occupations), 5 percent choose to be what I call tinkers (architects, engineers, scientists, mathematicians, and computer-related professions), less than 1 percent choose to be doctors (including physicians and surgeons, dentists, veterinarians, pharmacists, optometrists, and podiatrists), less than 1 percent choose to be lawyers (or judges or magistrates), and less than 1 percent are chiefs (CEOs).[144]

As you may surmise, the reason more people don't choose these occupations has a lot to do with required effort and time, individual interests, and individual ability,[145] as well as practical matters such as expense (education is expensive as you add up college and graduate programs and opportunity costs of years of lost labor). I believe the lack of supply for these careers also has something to do with what is *popular* in our culture and the choices we make when we are young. Math and science historically are *unpopular* subjects in school that geeks and nerds prefer. But as Bill Gates famously said (and he is at the top of the pyramid and the end of the income curve), "Be nice to nerds. Chances are you'll end up working for one."[146] In fact, based on the previous graph, the top 20 percent of our country is made up mostly of these very same nerds, and the other 80 percent are mostly the more *popular* people!

In part, the differences in these outcomes also have to do with differences in our socioeconomic opportunities, as suggested by our current president

---

[144] I used data from the US Census Bureau. I did not include PhDs as *doctors* here since the census data does not distinguish between scientists or economists as PhD versus non-doctoral degrees. I included these all more logically in the category of tinkers (a.k.a., thinkers). For doctors, I included all professions with known doctoral degrees, such as physicians and surgeons, pharmacists, doctors of optometry and podiatry, veterinarians, and all dental specialties. You could slice it up differently, but you will get the same skewed results, which is the main idea I am after.

[145] In this example, nineteen of these twenty occupations (95 percent) require at least a bachelor's degree (airline air traffic controller may get by with an associate's degree only). Six out of twenty require a master's degree as minimum entry level, and nine out of twenty are doctorate or professional degrees, with a combined fifteen out of twenty therefore requiring more than a bachelor's level degree. If you desire income as your outcome, get a higher level of education!

[146] Famous Quotes About, http://www.famousquotesabout.com/quote/Be-nice-to-nerds/65830.

(chief), which once more relates to our culture and what it choose to value and promote. Not everyone has the same opportunities as I have enjoyed, and this is certainly a function of your family of origin and its socioeconomics, your family and peer culture, and finally your own choices with the abilities you have to work with. This all seems so random, which is one of the reasons I am so grateful for my own opportunities to be able to do something so disproportionately rewarding. Regardless of what you attribute the outcome to, the imbalances still obey the same distributions, and these principles still apply—always and everywhere!

Most of these careers I have shown in this analysis by no coincidence require higher levels of educational achievement, more than simply a high school diploma. Less is not sufficient in this case. More education is in fact needed if you want to be in this elite group of income earners.[147] Most of these career choices are in fact education-intense and very competitive and involve more labor, more quality in education, more individual effort, and often more overall time to get the education, as well as more achievement. But the idea is still the same—a few occupations are efficient earners of income and predictably so. And if you learn this, you can be efficient in the allocation of your time and efforts. This is not a prerequisite for success and happiness, but it is certainly one factor that weighs in disproportionately.

Relative to the eight hundred or so occupations categorized by the Bureau of Labor and Statistics, it is important to realize that a vital few careers consistently do well for achieving decent income. Therefore if you truly wish to be efficient in your return on investment (ROI) with your career type (and your time or expense for your education), pick an occupation that is appropriately rewarded *and* in an area that you enjoy, and get the

[147] In fact, you start to diminish your efficiency the higher up you go in educational achievement, which is part of the dilemma. Examining average income per year of investment can be enlightening. For example, a physician earns a lifetime average of $6 million but takes twenty-eight years on average to get there in education (while accumulating some serious debt). A CEO earns the same lifetime average but only takes twenty-four years on average to do so, which is more efficient. Likewise a lawyer earns only $4.2 million with twenty-four years of schooling, which translates to less efficiency in earnings per invested time (if you don't count the return on the time later in career hours worked each week). Tinkers can earn $3 million in lifetime earnings with only a bachelor's degree, which is twenty years of education, and with less debt for a better ROI than the other choices in my opinion.

necessary education up front while it is easy[148] (and more affordable) and you are still young!

I could suggest also some internal efficiency here within the system with laws of supply and demand. Since less than 1 percent of the population becomes the doctor, lawyer, or CEO, why should these high achievers not naturally also be in the top 1 percent of income earners as they are presently?[149] And tinkers, as I refer to them, being three to five times more common (depending on what you classify as a tinker) should roughly make proportionately less income as the top income earners with roughly the same education, while merchants and managers, being even more common, should earn even less income. I refer again to a natural imbalance within our country's educational attainment in part 2 of this book, but for now suffice it to say we clearly have a mismatch between our educational desires and our educational achievements. Educational achievement is not popular within America, and it certainly is not popular in my uneducated state, as I have witnessed personally![150]

For my own curiosity and to see how the rest of the American population fares with income as a function of occupation (in order to better observe any disparity), I decided next to look at every American occupation (rather than as a larger industry grouping) and not just the top twenty earners, who can be unusually lopsided, like everything else. I was interested in the

---

[148] We learn more easily when we are young. I also believe getting an education is easier when your parents help offset the expense than if you wait and try later in life on your own dollar. Minimize costs, minimize your time to graduation, and match efficiently to the careers with little supply and much demand.

[149] The true earnings are, of course, hard to know, but the average numbers are roughly $171,000 for doctors, $120,000 for lawyers, $80,000 for engineers and architects, and $107,000 for managers (CEOs are $176,000). Engineering and architectural managers as a combo are also very rewarding ($130,000 on average), as a friend of mine would agree. Notice these differences are not huge, especially after you remove taxes as Uncle Sam always does.

[150] I talk about this again later, but West Virginia has the lowest post–high school educational-attainment rates in America over the last twenty-five years and probably even before that. And this trend will continue long into our future as well since the growth of educational achievement is proportional to the existing population rate of already low educational attainment. Unless we somehow educate more efficiently, our state will continue to grow at a similar rate and always be dead last in the United States, or the statistics could worsen if our youth leave the state to get educated. So in a way, we can once more "blame" this on our parents and grandparents who chose to live in and contribute to an area of uneducation. Do you now get what I mean when I say you can't pick your parents?

entire occupation-related income distribution rather than just for the elite few (the top 1 percent). So I further analyzed collectively all the categories of occupations for the entire US population. This was especially helpful for me to show my four children, since none of them plans to be in the top twenty occupations—not yet anyway!

As I stated previously, the Bureau of Labor and Statistics has summarily categorized over eight hundred different descriptors or groupings, which includes most American occupations. This list of professions by no means is all-inclusive and so does have some inherent flaws as I have noted. But it is nevertheless a starting point, and it has a lot of useful information that highlights the skewed nature of careers within America (using income here as the outcome measure).[151]

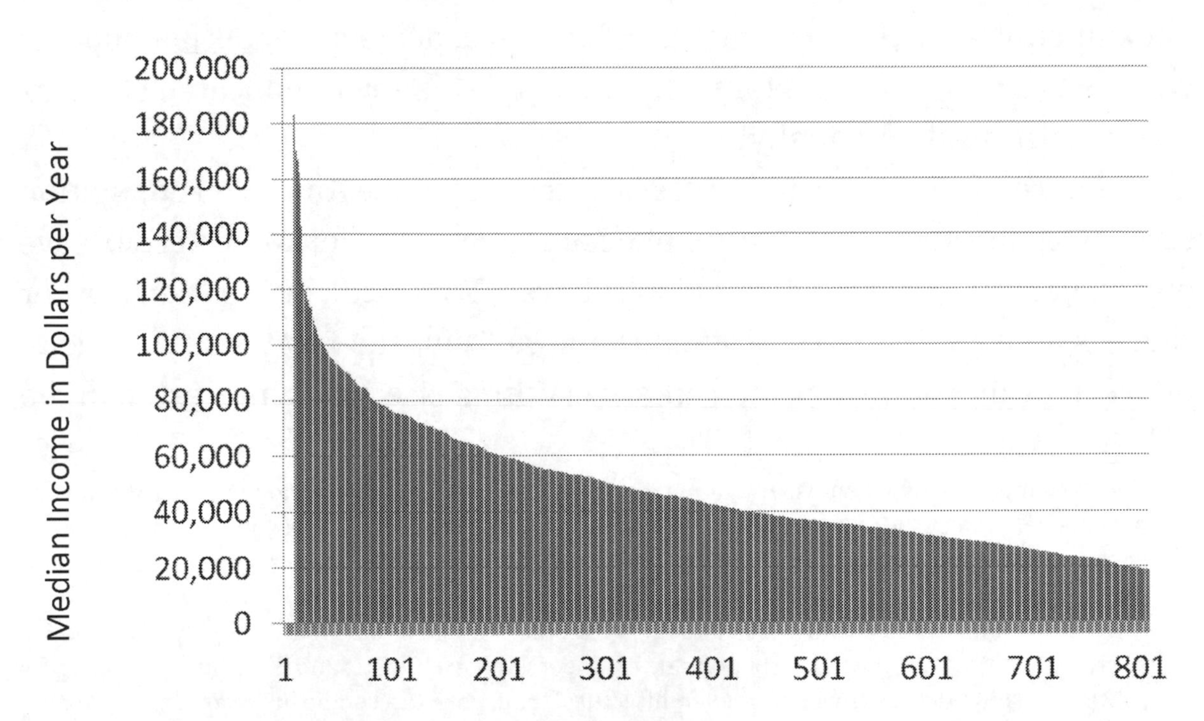

[151] I have similarly shown the same relationship of careers in America using population as the measured outcome to show what is popular in America (see images 4A and 5A in the appendix), which is an interesting lesson in itself.

This figure shows all occupations within America as listed by the US Bureau of Labor and Statistics (more than eight hundred) ranked from highest paying to lowest (by median income in 2011 dollars). As you can see, it forms a skewed distribution as well (mostly an exponential decay). A few occupations (the first one hundred or so) account for the high-income earners, and the rest of the population (85–90 percent) is at or below $80,000–$90,000 per year of occupational income (that is why I used that particular threshold a few pages back). And even within the skew, the top twenty look better than the rest of the occupations. The point is the skew, and that is all. A few occupations (less) on average pay more in the long run, and a lot do not pay much at all when you measure income as your desired outcome. This is similar to the distribution of household income that I showed previously, but with occupations the distribution appears less skewed and more consistent with simple supply-and-demand concepts.[152]

Occupational choices and careers are therefore not assorted or distributed normally or fairly in terms of incomes and earnings. Certain careers (about 10–20 percent) are favored (either by society or otherwise) over others, and the result is skewed as you can see from the graph. The career choices on the far right historically struggle, and the ones on the far left do not (an example of a struggler would be the tailor, which my first daughter plans to be). An efficient-minded person who is unsure of his or her career choice might consider these imbalances before spending a lot of time and money without a clear occupational or educational goal in mind.

When recently polled about this issue, many college students stated they were now chosing their majors and their occupations based on the

---

[152] I show the supply-and-demand relationship more thoroughly in the appendix. This graph above I obtained by simply ranking all eight-hundred-plus occupations in order and then graphing to show the imbalance is not as skewed as other distributions (it is not a power law), so I don't think occupational rewards (i.e., incomes) are an area we need to focus our efforts in order to improve our society. Household income that I show elsewhere includes multiple earners (cohabiting or married) and tends to show geographic trends and thus tends to magnify the effects of the differences we choose when we become households, in my opinion. However, we could lower this spread by providing more supply (hence education or importing these folks from other countries) and lessening the demand.

anticipated incomes from their occupations more than any other factor.[153] While I think this idea is in theory good (and yes, more efficient), it could really be a bad way of thinking if you choose your career based solely on income and then the predicted future income changes based on supply-and-demand economics or for some other unforeseeable reason. I would not choose engineering as an occupation, for example, if I did not have a passion for the analytical thinking or a propensity for math and science. Similarly I would never suggest picking medicine as a career in order to be a "rich man" unless you had a true passion for it, because it will consume a disproportionate amount of your time (more so than many other professions).

My children will hopefully learn to skillfully match their desires and passions with their abilities and then understand these skewed principles of outcome (especially with regard to use of time) as well. If they can learn to do this and achieve at their true passion, then happiness will follow naturally as the ultimate outcome, as will income and all the other measures I describe in this book.

I personally think it helps from an efficiency standpoint to match something we all enjoy *as well as* something that society appropriately rewards if we wish to remain happy and be able to support a family. By matching our passions and our abilities with the outcomes that society prefers as well, we should be able to find a nice compromise, a win-win decision.

Another idea I wish to emphasize here (especially for my children while they are young) is the generally repeating disproportional weighting of mathematics and technology/sciences in modern professions that consistently perform above the average. Occupations utilizing computers, life and physical and social sciences, engineering, and health care are all founded in science, technology, engineering, and math (so-called STEM

[153] William G. Bowen et al., *Crossing the Finish Line* (Princeton: Princeton University, 2009). This book appropriately focuses on the heart of the problem, which is trying to improve efficiency within our education system and get more of our college kids to complete college and graduate. Currently we are about 50 percent efficient in that measure. I am focusing on the "mind" of the problem, which is to change the way we value it within our culture. Changing that will help more disproportionately than anything our government does or that throwing money at the problem does (we already spend more per capita than any other nation in the world on our education).

areas), and they always seem to get rewarded quite efficiently. Business and finance and managerial professions made up the majority of the remaining high-income performers. While I am neither an economist nor a guidance counselor, I am almost certain these will remain good jobs into America's future, and I advocate these careers for their predictably imbalanced results. Stated another way, if you have any inclination in math or science, I suggest capitalizing on that strength when choosing a future occupation. These are the most-efficient fuels for a high-tech engine within modern America, and this will continue more so into our future.

Supply-and-Demand Equilibria and the Value of Thinkers

Pareto similarly noted the efficiency of the law of supply and demand one hundred years ago in *The Mind and Society* treatise. "Velocity in circulation (among the social classes)[154] has to be considered not only absolutely but also in relation to the supply of and the demand for certain social elements ... In a country where there is little industry and little commerce, the supply of individuals possessing in high degree the qualities requisite for those types of activity exceeds the demand. Then industry and commerce develop and the supply, though remaining the same, no longer meets the demand."[155]

In fact, the curves for supply and demand have origins from what are known as *indifference curves.* Pareto was familiar with these curves and referenced them. They are very similar in shape to Pareto's curve of income imbalance and likely influenced him.[156]

Lastly, the final point I wish to make here relates to the idea of quality and achievement relative to others within your same occupation. If you

---

[154] Note the relativity of all of this—velocity or movement is always a change in position of one thing to another over time. This is therefore all a measure of how we perceive our change in social structure compared to others, and right now in society, the top 1 percent and above are accelerating away from the rest of society. Velocity is usually first-order math, and acceleration is second-order math (a square function) like gravity.

[155] Pareto, *The Mind and Society,* 3:1426–27.

[156] F. Y. Edgeworth was a lawyer, who had a fascination for math and wrote a paper on the mathematics of psychology in 1881, *Mathematical Psychics: An Essay on the Application of Mathematics to the Moral Sciences.*

are a person who likes to be above average in whatever you choose to do, you could perhaps expect higher relative earnings than the average (or the median), and I show this range in income (percentiles) below for the various industries just to give you an idea of how quality and achievement can factor into your occupation as well, with some occupations demonstrating this more than others.[157]

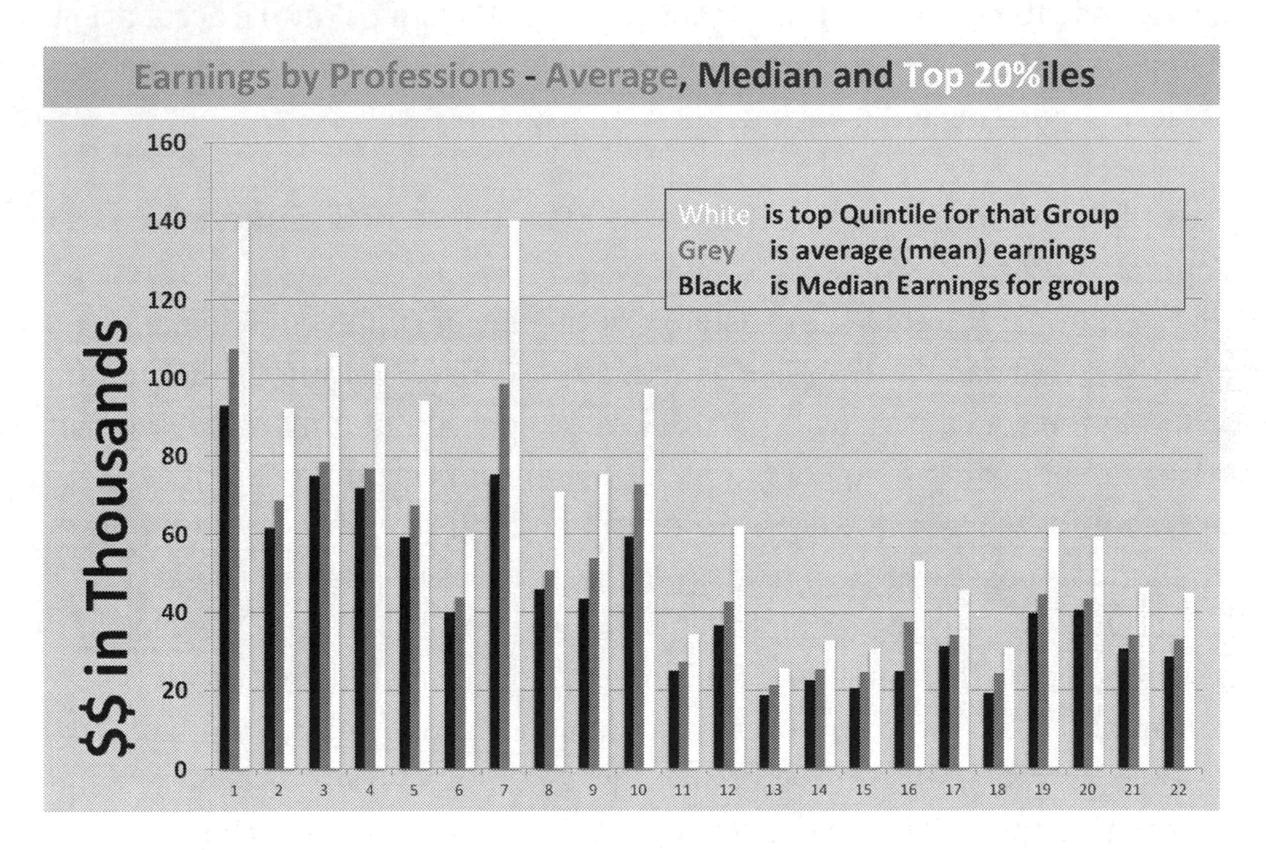

All Occupations/ Professions by Groupings as listed by US Census

Here I demonstrate the spread of income within various occupational groups, but instead of using just the average or median income performers, I show higher levels of achievement. As an average achiever you might

[157] On the flip side, the inefficient performers within each category could be examined as well, and I show this in image 8A in the appendix (I show the bottom 10 percent versus the top 10 percent for each occupational category). I could say quality is a function of geography and many other variables and not entirely intrinsic to the person since I don't have a way to analyze quality here, but I hope you get the point. There are ranges for each occupation that are also skewed, some more than others and some less.

expect the average (or the median) earnings. As an overachiever or high performer, you could reasonably expect these higher ranges (such as the top 20 percent) for that particular occupation. I show this information to emphasize another point here. Certain occupations don't ever seem to do well for income potential, no matter how well you achieve (e.g., food preparation, grounds and maintenance and repair, personal care, and farming, as seen in the above percentiles). You may also note the occupations that have the highest disparity between the median earners and the top 20 percent (like management and legal professions), although this would be better examined looking at the bottom 10 percent relative to the top 10 percent.[158]

Based on this occupational information that seems to repeat regardless of the time period examined, I would still encourage my children over the next twenty years to efficiently entertain careers in health, medicine, or dentistry (the doctor); law (the lawyer); marketing, finance, and business (the merchant); science, math, and engineering (the tinker); computers, programming, and software (also the tinker); management; and as CEOs (the chief). The majority of the occupations in America that pay well are industries that rely on you using your brain disproportionately more than your brawn. These are all occupations therefore that are made up collectively of critical or vital *thinkers*. All these occupations that pay disproportionately well, as well as the ones that don't pay well, follow the principle of imbalance, or you can think of it more simply as the related laws of supply and demand.

I personally believe this data also illustrates the cultural fact that industrialization of our world has shifted society from one that relies on

[158] These are the twenty-two groupings used by the Census Bureau to categorize the various industries (in this graph my numbers correspond to the following categories): 1=management, 2=business and finance, 3=computer and math, 4=architecture and engineering, 5=life, physical, and social sciences, 6=community and social services, 7=legal, 8=education, training, and library, 9=arts, design, entertainment, sports, and media, 10=health-care practitioners and technical operations, 11=health-care support, 12=protective service, 13=food prep and serving, 14=building, grounds, and maintenance, 15=personal care and service, 16=sales and related, 17=office and administrative support, 18=farming, fishing, and forestry, 19=construction and extraction, 20=installation, maintenance, and repair, 21=production, 22=transportation and material moving.

human *physical* labor to one that relies on *mental* labor (globalization has simply magnified the effect) and thinking. While we need both, our ability to critically analyze and solve problems, much like Pareto showed in his accomplishments, is what we are gravitating toward in our world of increasing specialization, as is evident currently within America. Even our country's name derivation hints at this hidden force of nature at work beneath the fabric of our own work ethic (*Americus* means "industrious leader").[159]

In our exponentially growing world of technology, as we continue to develop machines and tools that replace the manual labor of many men with a few more-efficient machines, the need for physical skills diminishes disproportionately,[160] and the result is the relationship that we now see prevalent in American pop culture.

Stated more simply, we have evolved[161] into a new world that disproportionately values thinkers, and that is simply a modern reality in a universe that continues to disproportionately reward efficiency.

In part 2 of this book, I show how understanding this same imbalanced relationship applies to the more general concept of human choice and to the idea of the mechanical efficiency of our efforts as described by Harvard educator George Kingsley Zipf.

---

159 America derives its name not from Italian Columbus (Cristoforo Columbo), to whom we attribute the geographic discovery of America in popular culture, but from another Italian *merchant* and explorer who explored the Americas (mostly South America), Amerigo Vespucci. *America* is the feminized version of "land of Amerigo." The name *Amerigo* in Italian (from the Latin *Americus*) means "industrious leader," and supposedly Vespucci was named after the virtuous Saint Emeric, which is where *Americus* is derived. So whether you realize it or not, our country's name (which is after a merchant) embodies these very ideals of industry, elite leadership, and religious virtue (i.e., the very elite to which Pareto referred).
I would be derelict not to mention Spain's role as provider of the Catholic monarchs and thus the finances for the Spanish colonization of the Americas by employing Italian merchants known for their vessels' sailing efficiency and quality, as well as Columbus's primary motivation of spreading Christianity to nonbelievers. *Wikipedia*, s.v. "Amerigo Vespucci," last modified August 10, 2014, http://en.wikipedia.org/wiki/Amerigo_vespucci.

160 I also included two additional curves in the appendix for these occupational outcomes by (1) income and industry and (2) popularity and industry to show the same exponential relationship with demand (what we are willing to pay) and supply (what we are choosing to do). They parallel each other nicely, which suggests an efficient supply-demand relationship. See images 4A and 5A in the appendix.

161 Notice the nod here to Darwinism and the idea of evolution as a means of efficiency as well as how *ideas* evolve similarly in our culture.

PART 2

Zipf and the Principle of Least Effort

"Grandmother, Grandmother, What Should I Wear? Silk, Satin, Cotton, Rags."

Communication Is Imbalanced in Predictable Ways

As this project evolved for me, I began to see a unifying theme emerge, like some hidden fabric in space and time weaving its skewed way into my field of vision. Whether the skewed imbalances I noticed ubiqitously involved nonhuman physical events within the grand universe or how we tiny, insignificant humans measure our happiness (like what my daughter chooses to wear or how we can all shift our outcomes from *rags* to *silk*[162] [i.e., riches]), I discovered weirdly similar and skewed relationships everywhere, as if we were all obeying the same physical forces[163] of creation.

Whether it's the efficiencies in our choice of occupations based on supply-and-demand rules or more simply how we choose to attract things we value as a family, including how we choose to name our kids,[164] the outcomes seem repeatedly imbalanced, following similar rules of efficiencies (or inefficiencies depending on your frame of reference[165]) everywhere.

---

[162] One version of this rhyme uses *calico* instead of *cotton.*

[163] I would be derelict if I did not point out to you there are only a few (four) such universal forces that make everything else happen (gravitational force, weak force, electromagnetic force, and strong force). They obey power-law principles as well and are scale invariant (independent of how you measure them in space or in time).

[164] Names are based on popular culture and often reflect religious themes (and secular leaders in the case of Romans) historically, which all have to do with how we are patterned to think through our cultural influences. My name, Paul, is based on the biblical reference, as are many names. The point is the patterns are predictable and based on our culture and what is valued and/or popular.

[165] If you were on the planet farthest from the sun, you would conclude gravity and other forces were not in your favor, and thus you might think they were inefficient toward your well-being. If you were not educated well, you may think the system is inefficient for you; if you were on the low end of the income scale, you may likewise conclude society was inefficient at allocating its distribution of income. It all depends to some degree therefore on your perspective relative to the curve.

It now seems quite obvious in hindsight that these outcomes are skewed this way, since we all naturally differ in our desires and in our abilities, but what became more fascinating to me was the predictability of it all and just how skewed the results can be. I pefer predictable. In theory, I can even fluorish and be happy in a world that is highly predictable and become more efficient[166] by adapting to it.[167] The fact that these rules can be enumerated in a way that allows me to predict my own outcomes and those of others truly amazes me.

As I've stated in previous chapters, the principle of imbalance, as described so methodically by Pareto, is a *power-law relationship* that is repeatedly noted within all nations and from all time periods. We see this same phenomenon within the world of retail sales, Internet popularity, and anything that involves human preference (or attachment), often relating to the concept of economics (or as I discovered, religion, education, family values, etc.). But more peculiarly, the skewed phenomenon also becomes periodically evident within the nonhuman realms of mother nature. So I next began to examine the idea of these power laws to better understand their unique nature.

Power laws are not well known to most of us, although they are in fact ubiquitous. I was not aware of them at all before researching these concepts (at least not in name, although admittedly I knew about their behaviors). In power-law relationships, a result or outcome measure is often *a power* of some related variable (which I call an input). It is not a linear relation, therefore, and the results are usually way *more* (or significantly *less*) than you would otherwise predict.

---

[166] I suggest that we may have to give up some of our *efficiencies* if we desire more equality in society. Being more inefficient may actually help lower disparity and paradoxically increase general well-being. I would also be derelict if I did not point out that our ability to predict outcomes refers once more back to Pareto's premise that all these abilities we have are imbalanced in the same way. Some people are better able to predict outcomes by understanding these rules.

[167] If this sounds remotely Darwinian, it would be no coincidence. His theory on evolution was similarly based on these same ideas of efficiency and adaptation. One model known as the Yule-Simon distribution (which is Pareto-like) explains evolution similarly. Others, including Fascist thinkers and Italian Mussolini, in fact seized on Pareto's ideas' similarity to a socioeconomic *survival of the fittest*, which was not Pareto's original intent, although he was likely swayed by the prevailing geopolitical Fascist culture of Italy when this did occur in his country (pre–World War I).

For example, my body's basal metabolic rate (the energy needed for me to stay alive) is a function of my body mass raised to the power of 75 percent or 0.75.[168] This is a weak power law since the power is less than one, but nonetheless an important relationship to know. A person with a larger body mass requires and expends disproportionately more calories of intake (input or food) to function. This seems logical and has everything to do with physiological efficiency, but what is important is the significance of the relationship[169] and the fact that one variable is a power of the other and therefore not just a simple linear relationship.

When I performed a general informational search for *power laws*, I found a few common examples that at first glance seemed rather uninteresting. Earthquake sizes, for example, follow these empirical rules,[170] as do craters on the moon (their formation from the collisions of ateroids), solar flares, and other related ideas (e.g., initial mass function of stars, energy spectra of cosmic ray nuclei, and Kepler's law of planetary motion, which later led to Newton's law of gravity[171]). On earth, the foraging patterns of animals follow these laws nicely, as do scaling (growth) properties of plants. In biology, neuronal activity (our brain behavior) follows their mechanics, as does the growth of many of our bodily organs (a concept known in biology

---

168 This is known as *Kleiber's law* and is a power-law relationship seen in biology. Max Kleiber, "Body Size and Metabolism," *Hilgardia* 6, no. 11 (1932): 315–51. Other examples include the more general *principle of allometry*, which relates body size (our mass) to our shape, anatomy, and physiology. Our brain neuronal development and vascular organs and lymph system, for example, follow fractal (power laws) mathematics that relate to this same concept, which developed out of these same relationships (fractals evolved from a mathematician at IBM named Benoit Mandelbrot, who in fact read the work of Zipf). Power laws are universal tools working behind the scenes to guide our growth and development in ways we are only beginning to understand. Amazingly, I had never heard about these biological laws before—even as a physician!

169 This is especially relevant if you are trying to be healthy and want to calculate your daily intake to avoid obesity or if you are bodybuilding and want to calculate the necessary calories needed to efficiently add *lean* body mass without too much additional fat.

170 Gutenberg-Richter power laws are one example. An excellent book on this subject matter (and mostly about power laws) is Mark Buchanan's *Ubiquity: Why Catastrophes Happen* (New York: Crown Publishing, 2002).

171 Were it not for these power-law distributions, we would not in fact exist. They explain the skewed imbalance of everything in the universe from dark matter to human DNA. Gravity is essentially the main force of attraction in the universe, especially when we get to large distances. On smaller scales, electromagnetic forces become more prevalent (which are both attractive and repulsive forces depending on electrical charge and magnetic poles).

as *allometric* growth,[172] which interests me a bit more than craters on the moon). The sizes of power outages and the deaths that occur from war follow these same weirdly predictable outcomes, much like our findings of income imbalance. Weird, huh?

For fun, I created a graphic of a few of the more common areas where scientists have noted power laws as repeating mathematical relationships. These are shown below for the reader's interest, but to me it hints at some universal link to *a rhyme and reason* within our creation and for our existence (remember what I said earlier about God and dice?).

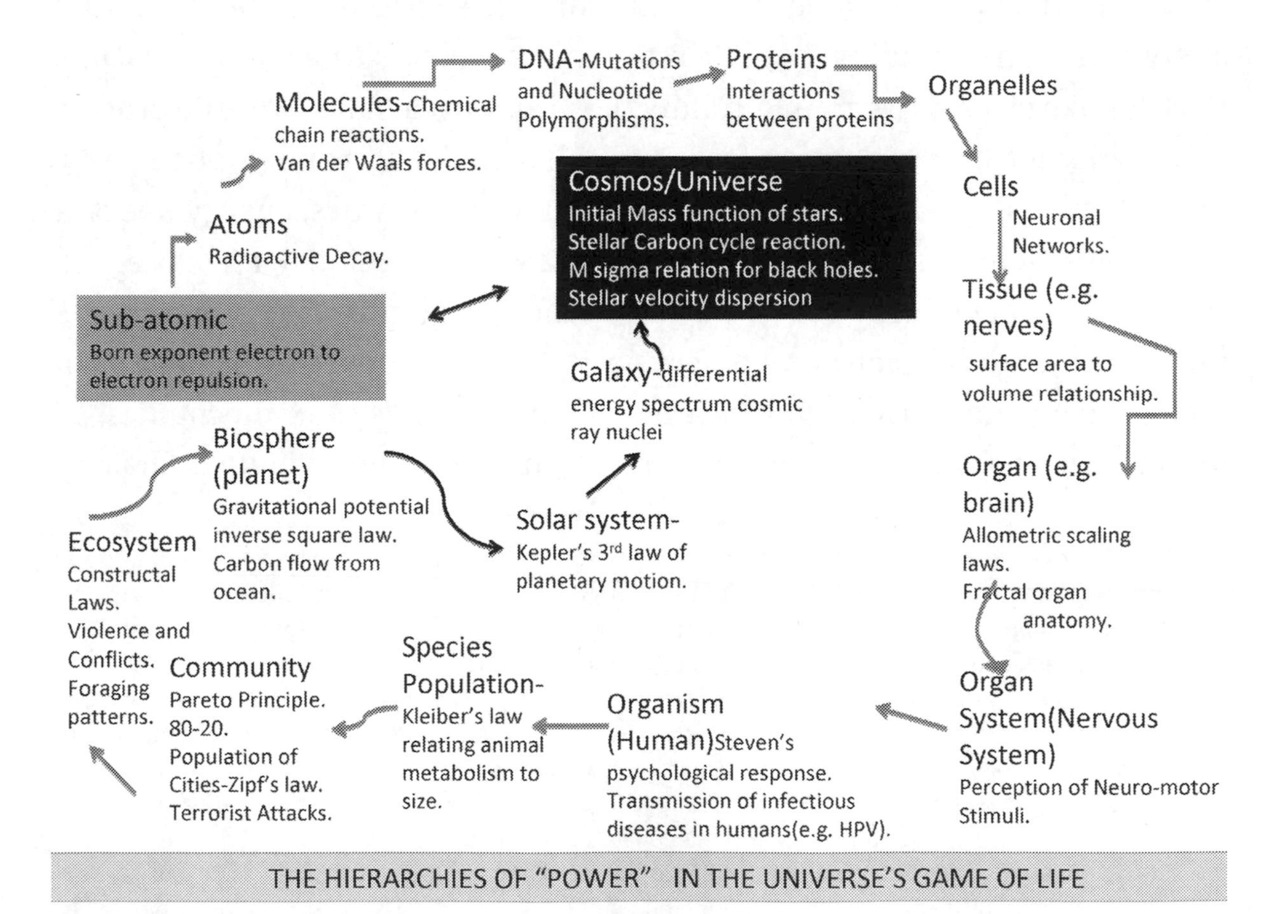

THE HIERARCHIES OF "POWER" IN THE UNIVERSE'S GAME OF LIFE

[172] For example, our digits and our skeleton grow as a power of our mass to conserve energy expenditure and optimize efficiency; our metabolic rate is a power function of our mass, as is our heart rate (to optimize efficiency in delivery of nutrients and blood to the body parts). The point is simply that these forces are acting on our bodies to create these efficiencies whether we know it or not! Thomas T. Samaras, *Human Body Size and the Laws of Scaling: Physiological, Performance, Growth, Longevity and Ecological Ramifications* (Hauppage (NY): Nova Science Publishers, 2007).

Here I have displayed a few areas of known power-law relationships that are visible within the greater universe as well as some invisible ones that weave their way into our own (human) natural imbalance through forces of attraction (like gravity, which is shown above as well). These are all simply scaled versions of the same ideas.

Experts currently believe the universe began as a big bang (either by godly intervention, purely by scientific principles, or—as I believe—both[173]). At some instantaneous moment in early time (I believe time can also be conceptualized as nonlinear as a result of this very idea of imbalance), one element, hydrogen, disproportionately accounted for 74 percent of all of the matter and energy in the universe while another element, helium, accounted for the other 26 percent.[174] Everything else did not really exist until the power of time mixed with three to four other vital forces.

---

173 I prefer to accept multiple theories (win-win) rather than ignorantly presume one is right and all others are wrong (win-lose), because there is no way we could ever know right or wrong. I am all about balance. We will continue to learn more as time goes on, but I feel we are limited in how quickly we can learn, and thus it will likely take many more years before we figure all this out. I prefer in modern thinking to have a single beginning from whence everything else emanated, whether that is an anthropomorphic God or simply mathematical rules of explanations. But I also prefer to think there is a guiding wisdom rather than simpler amoral law and valueless principles. If there is no moral or ethical master other than the common conscience, then we can mostly choose to do anything we please (including all degrees of negative behavior) as long as our fellow humans don't figure it out and penalize us. I personally prefer a God with purpose and morals who exercises His will through nature and through these universal principles that we must sort through at our own levels of efficiency. The sooner you figure it out, the quicker you get to the next level of the hierarchal (and universal) game of life. It is similar to Maslow's pyramid concept of social development.

174 Lawrence M. Krauss, *A Universe from Nothing* (New York: Free Press, 2012), 114. The distribution of matter in the universe data I use here is as predicted scientifically at the time of nucleosynthesis (i.e., the universe's birth). A universe that continues to exist from a theoretical combination within infinite multidimensional universes is one that is thermodynamically or otherwise stable (e.g., quantum gravity allows an expansive universe, which favors a directional arrow of time that then allows us to exist due to the second law of thermodynamics, which allows time directionality so events have a future effect linked to a past cause or choice, creating a consequence relationship as I have shown). Philosophically it does not matter so much why or how; it just is what it is. But if you believe in directionality and a starting point (like a big bang), you likely find it appealing to think our behavior is guided and influenced by either sound principles (i.e., forces) in the universe or by a deity that enabled these forces from the start and continues to direct them. To such a force or deity, time and space become irrelevant concepts (and only relative), although from the perspective of the

Over time and thanks mostly to gravity and strong (and weak) nuclear forces, the universe became familiarly skewed to current elemental distribution: 70 percent hydrogen, 29 percent helium, and 1 percent *rare elements*. This *elite* (1 percent) rare elemental matter includes oxygen (which accounts magically for 20 percent of our atmospheric air which seems critical to our existence) and accounts for 80 percent of our planetary elements.[175] Complex chemical compounds are derived from this elemental matter, including proteins, lipids, sugars, and so on down to the superefficient construction of and reproduction of DNA,[176] to our cellular communities, bodily organs, and then into our human machine of continual networked efficiencies. If we were not so efficient from the perspective of time and these forces of nature (whether or not you call this evolution does not matter), we simply would not exist. These same rules have gotten us to where we are with the principle of *more from less* ad infinitum.

So my next questions were more simply, "Are the societal outcomes (e.g., within America) all just human equivalents of natural physical principles (like the force of gravity) of matter interacting with other matter? And if so, can all of these outcomes likewise be predicted with statistical modeling?" Well, Pareto seemed to think so,[177] and so did a few other thinkers, including German American professor, Dr. George Zipf.

Interestingly, the best data I could find on the nature of power laws and their weirdly repeating skewed distributions came from Dr. George Zipf, a man who studied *words* among other things in the early part of the twentieth century. The frequency of words in most languages fits a repeated

---

observer these concepts mean everything; the space and time differences by definition constitute our miniscule lives.

175 This is an example of what is known as the 80-20 rule.

176 DNA codons are distributed similarly by what are known as Yule distributions, and in fact Zipf's law may be a specialized version of the more generic case of a Yule frequency distribution where there are fewer entities involved from which to choose. Yuri Tambovtsev and Colin Martindale, "Yule Distribution and Phoneme Frequencies," *Journal of Theoretical Linguistics* 4, no. 2 (2007): 1–11.

177 Pareto used Newton as a reference in establishing the laws of sociology, referring to the same ideas Newton used to establish the scientific method and the law of gravity. I guess I could more simply argue we are just continuing to think about all these concepts in the same way, as learned from those before us, but it seems to be more than that. There is an inherent order that seems to be common regardless of what words or numbers we use to describe it.

pattern that reveals the very nature of the principle of imbalance in regard to communication. Dr. Zipf was a philologist,[178] a man interested in language through literature, and while teaching language at Harvard for practical reasons, he pursued his true passion with his research into the mechanics of human language. His results had much broader applications to general human nature than simply to our mundane choice of common word use.

George Zipf observed that a few words are used very often in human[179] languages (they are popular by choice), while most words are used rarely. Zipf found this imbalanced relationship surprisingly predictable in speech and human literature, and he quantified it. In fact, word use followed a frequency of inverse proportion with relation to the words' utility.[180] In other words, the most popular choice of a word was used most commonly (and would be number one in rank at some natural frequency), the second most common word would occur about half as often in relation to the first (50 percent relatively), the third most common word one-third (33 percent) as often, and so on until each word was used once only. This is now referred to as *Zipf's law* and is actually another example of a power-law relationship. The important idea here is to understand the relative nature of one (in rank relation) to the others. As it turns out, we as humans similarly rank everything we do in life by a similar internal comparison of ourselves to others.[181]

---

178 *Philology* originally came from Greek (and later Latin and our Romans once more) and historically means "love of knowledge" but later morphed into "love of language" and then finally a "love of language through literature." Zipf loved it all—but mostly knowledge. It is interesting how words evolve like people and cultures to take on *more*-relevant meanings *for* the times and have *less* of their original intent.

179 After Zipf's death, this was also shown to be the case for animals such as dolphins and is likely the case for any species that communicates with some internal efficiency using a receiver-transmitter relationship.

180 A rank-size distribution often is called Zipfian since it obeys these rules. So many of the things we value in society and rank follow these same principles. Generally speaking, when you rank many variables and they seem to follow this relation, it is likely Zipfian as well.

181 This idea of comparison to others is what makes income disparity and other disparities so important. As Pareto said, "A word designates a concept, and the concept may or may not correspond to a thing. But the correspondence, when it is there, cannot be perfect. Even if the word corresponds to a thing, it can never correspond to it exactly, in an absolute manner. *It is always a question of a more or a less* ... In a word, the 'absolute' has no place in [logico-experimental] science, and we must always take in a relative sense propositions that in the dress of ordinary parlance seem absolute; and in the same way too, we must make quantitative distinctions where common speech stops at the

Let me reveal to you a relevant example of this hidden (Zipf's) law to demonstrate how I believe it invisibly guides our behavior and our efficiency through communication. To do so, and if you are so inclined, go back and read again the simple counting nursery rhyme from the preface.

Counting the words used in the nursery rhyme "Tinker Tailor," I discovered that, although short in content, its word content follows Zipf's law. *Shall* is used seven times in the entire rhyme and is the most commonly repeated word. The second most common word is used half as frequently (four times); the third most common word is employed one-third as often as the first, and so on in a numerical frequency by rank that is both rhythmic and mathematically predictable.[182] Most words are in fact trivially used only one time while a vital few words are repeated more often. Is this a coincidence? No! *There are no coincidences*, at least not in the sense you are thinking, which is some purely random chance. Does it matter? Yes, because it suggests an underlying repetitious process (language in this case) that helps demonstrate the predictability of most human behavior and efficiency! We are internally programmed this way in everything we do.

Word use and our ability to communicate is where I first discovered power laws' practicality, as pointed out by Dr. Zipf. Most words in most languages are in fact seldom used in communication, and a vital few words are used more frequently to help communicate the intended message (in the case of the nursery rhyme, the message is about choices and consequences).

---

qualitative ... to express ourselves always with absolute exactness would be to wallow in lengthy verbosities as useless as they would be pedantic." Pareto, *The Mind and Society*, 1:57–58. This sounds very much like what Zipf argues later in terms of efficiency in communication. The two men both agreed that mathematical interpretations of human behavior allowed more general developments of principles and laws rather than using the inefficiencies and ambiguities of words.

[182] Word use in terms of frequency and rank follows a type of mathematical distribution known as a harmonic series. Word use in frequency is inversely related to utility. If you count out word use, the first word in rank is the most commonly used, the second turns out to be used half as much, the third occurs at one-third of the frequency of the first, the fourth is used one-fourth as commonly, and so on. This is a special type of power law with a power of −1. I think of it as being easier for me to reuse a few words more often to convey my ideas rather than memorize a gazillion words that I only use one time. *My brain likes easier, and so does yours!* It has everything to do with efficiency, in this case with communication. Mathematically, the frequency of word use = $0.1\ r^{-1}$ where r is the rank. This is what is now referred to as a rank distribution.

In "Tinker Tailor," the three most common words in order of rank give you "Shall I Daisy?"

While to you this may seem a simple and mundane example of a force of human nature directing your use of everyday words, it is much more powerful and certainly more universal than that. These forces are *invisibly* directing more than just your use of word frequency, as I intend to show you! They are directing the very nature of how you think, as George Zipf showed, and these principles act by affecting the frequency of your word usage in order for you to communicate with others more efficiently—not just now but always and everywhere! It is even invisibly at work in shaping how we name our children.[183]

Have you ever heard of Zipf's law before?[184] Me neither! Not until I started this book. But this commonly unknown law relates to more than just word

---

[183] Names like mine (George) also follow similar outcomes of cultural popularity and often have religious themes. For example, my first name comes from Saint George, who was the product of two Christian Greeks during the Roman era. Saint George was executed for his beliefs and his failure to renounce his Christian beliefs in favor of the Roman pagan gods. Before the then-governing elite Emperor Diocletian had him executed, George gave away his wealth to the poor and was later tortured and beheaded on the twenty-third, which is coincidentally the date of my (and my son's) birth. Saint George is often depicted in art as a figure slaying a dragon, which symbolizes the battle of man against Satan. Diocletian's wife, Alexandra, is often depicted in these images as a maiden in the distance. It is ironic that I had no idea about any of this before looking it up recently. Coincidentally, my wife and I named our son Alexander (after Alexander III of Macedon, a.k.a. Alexander the Great), which we were prepared to change to Alexandra were we blessed with a girl. Interestingly, Saint George is one of a few saints and legends known and respected by Muslims as well as Christians. There can be a lot of power that is secretly hidden within a simple name or even in a single word.

[184] Zipf's law was originally used to describe the repeated pattern of word use, but Zipf later applied it to other areas as well. In the original description the law was used in an analysis of million-word documents to show which words were most commonly employed. Words like *the* and *of*, for example, typically account for the majority of words in a document and are distributed in a way that is inversely proportional to their frequency. The original context was a compilation of English text documents that were analyzed for many patterns in linguistics, psychology, and other measures. Interestingly the frequency distribution of word use was almost exactly the same I found in this nursery rhyme above: the most common word in the nursery rhyme was *shall*, which accounted for 8 percent of the total word count. In the Brown Corpus analysis the most common word was *the*, accounting for 7 percent of the over one million analyzed words. The next most common word was then about half of that, or 3–4 percent. For the rhyme *I* and *Daisy* tied for number two in rank at 4.5 percent each. In the Brown Corpus *to* and *of* tied at second with 3 percent each. In a larger study of many more words than I analyzed here, Zipf found half of the vocabulary came from words used only once (hapax legomena), which held true for the nursery rhyme (forty-five of the ninety words I counted were only used once). Pretty amazing that this is accurate in the Brown Corpus, something

use and frequency, as I will show you throughout this book. Zipf's law is just one example of universal efficiencies operating all around us, and it helps explain human behavior and why we naturally try to achieve *more for less*.

A subtle basic principle governing the common use of human language seems so simple in concept, and yet it is not widely known or taught in our educational systems or by our parents. Perhaps knowing it and the other power laws I have discovered would yield too much knowledge (about how life works or how *inefficient* our educators and parents are) and hence too much power! As I like to say, "Simple is elegant." And as Albert Einstein said, "When the answer is simple, God is speaking."[185] So you may want to pay attention here!

Zipf was methodical in his efforts. In fact he reminds me a lot of Pareto in his intentions to reveal some method (or some *rhyme or reason*) to his findings and to our madness. I could not in fact find much history about the man, since he died at the early age of forty-eight (coincidentally my age as I write this book), but what I did find were prolific writings from his work while at Harvard, which culminated in his book that elegantly described the mechanics of human behavior in 1949.[186] It is a fascinating read if you are so inclined, and I am certain his text obeys the same rules of word frequency and selection, were it so analyzed. Nevertheless, his work can be distilled to a few simple ideas I think are worth sharing.

Basically we are skewed as a culture, and we use as little effort (or work) as we have to in most everything we do, including how and what we choose to learn![187]

---

composed many hundreds of years ago in a different time and place. Even more amazing is that I tested it and it worked on the rhyme as well. Coincidence? No. The message that would be conveyed in this case would be "Shall I, Daisy?"

185 If you learn more about Einstein's life, you will find his view of God is not traditional (especially knowing he was born a Jew), but his reference is nonetheless reverent to a divine creator. Isaacson, *Einstein*, 391.

186 George Kingsley Zipf, *Human Behavior and the Principle of Least Effort* (Cambridge: Addison-Wesley Press, 1949). Within the body of his textbook, Zipf does in fact refer to the Pareto school of economics, which by this time had developed in Switzerland (Lausanne) where Pareto was a professor.

187 While this sounds bad, it really isn't—it is human efficiency as an organic being to conserve energy and waste as little as possible. I think of the analogy of a machine (like a car) and its inherent efficiency with some energy lost as heat instead of the desired outcome of mechanical energy in

Zipf believed that we could view human behavior as a natural phenomenon like everything else in the universe, with "fundamental principles that seem to govern important aspects of our behavior, both as individuals and as members of social groups."[188] According to Dr. Zipf, "Knowledge of these underlying principles will inevitably help others to live more efficiently, whether as individuals or as members of teams that co-operate and compete—and whether in the roles of those who primarily do the leading or in the roles of those who primarily do the following."[189]

He considered the human being like any other objective matter in the universe, acting in what it deemed the most efficient way to minimize its energy expenditure. According to Zipf, "Every individual's movement will always be over paths and will always tend to be governed by one single primary principle which, for the want of a better term, we shall call the *Principle of Least Effort*."[190]

To better define the principle, "A person in solving his probable future problems, as estimated by himself, ... will strive to solve his problems in such a way as to minimize the total work that he must expend in solving both his immediate problems and his probable future problems. That in turn, means that the person will strive to minimize the probable average rate of his work-expenditure (over time)."[191] The principle is "contingent upon the mentation of the individual," which in turn includes the operations of "comprehending" the "relevant" elements of the problem, of "assessing their probabilities," and of "solving the problem in terms of least effort."[192]

For Zipf, who was fascinated by word usage, words are merely *tools* used by humans to communicate a desire. For him, the study of words offered a key to understanding the entire speech process, and the study of the entire speech process offered a key to understanding the personality and entire field of biosocial dynamics. Human behavior can be modeled by speech

---

the form of wheel motion. Or think about a simple lightbulb that gives a small part of its released energy as desired light while most of the energy is wasted as heat.

188 Zipf, preface to *Human Behavior and the Principle of Least Effort,* v.

189 Ibid.

190 Ibid.

191 Ibid., 1.

192 Ibid., 7.

just like income and every other model I have shown you. Reread that last statement.

"Man talks in order to get something."[193] Hence man's speech may be likened to a set of tools that are engaged in achieving objectives. In fact, everything we do is to get something—whether that is talking (or texting, for my children), counting, singing, painting, educating, working, mating, shopping, writing, dreaming, thinking, or sleeping! It all boils down to this simple economic model of doing something to get something else (whatever the desire).

I have two points to emphasize here from Zipf's literary work. First, nothing comes without effort-you must do something to get something. Second, there is an implied interaction between two or more people that guides these relations (hence the idea of *relativity* and a *more or a less*).

Zipf eloquently describes his model as a dynamic system that links *tools* (in this case, words) and *jobs* (in this case, conveying meanings in order to achieve objectives).[194] "Each person develops their own set of *tools* (words) in order to convey their needs (to someone else), and this follows a relationship that has its own internal economy—there is the economy of the speaker (who tries to use as few words as needed to convey intent) and there is the economy of the listener (who tries to learn the meaning of what you are trying to convey). These form opposing forces that in the case of the speaker will tend to reduce the size of the vocabulary to as few words as needed and ideally one word with all meaning (a force of unification) and in the case of the listener will tend to increase the size of the vocabulary to a point where there will be a distinctly different word for each different meaning (a force of diversification). These two opposing forces merely describe two opposite courses of action, which from one viewpoint or the other are alike economically, and will from the combined viewpoint, be alike adopted in compromise. Whenever a person uses words to convey meaning he will automatically try to get his ideas across most efficiently by seeking a balance between the economy of the small wieldy vocabulary

---

193 Ibid., 19.

194 Ibid., 19–55.

of more general reference on the one hand and the economy of a larger one of more precise reference on the other with the result being a vocabulary balance between the forces of unification and diversification."[195]

Zipf, like Pareto, also became interested in applications of his principle to other realms of general human behavior. His work delved into subjects as diverse as city sizes and the concentration of economic power and social status.[196] His immense effort resulted in a similar treatise on human behavior, which I believe placed *more* emphasis on the mechanics of that behavior and the frame of reference (FOR)[197] and *less* on the actual outcomes themselves.

There is no question Zipf was intimately aware of Pareto's work, although I am not certain how much influence it had on his own theories. Pareto's work did disproportionately influence several other Harvard elite who imported his ideas to America by way of Harvard.[198] On several occasions, Zipf refers to the Pareto school of sociology and economics, as it became known from Lausanne where Pareto had chaired a department and taught in his final years. So while I am not entirely certain just how much these

---

[195] Ibid. The force of unification is analogous to the law of attraction (much like gravity), and the force of diversification similarly would be the law of repulsion (like electromagnetism with two of the same charges repelling each other).

[196] Zipf died young but did so much with his work in the years he was alive. It is truly amazing. His book is a wealth of knowledge. It was published in 1949 and was a vital and influential source of information for me in this book. His law that oversimplifies all the other relations he described relates to word frequency. It is described by the power-law relation $f(r) = 0.1/r$ or $0.1\ r^{-1}$. Thus the most common words we use in rank ($r = 1$) have a frequency of about 10 percent, the second most common words ($r=2$) have about 5 percent, third 3.3 percent, and so on. The tenth most common ($r=10$) would have 1 percent frequency.

[197] This is a link to my book title implying the need for us to each find our own frame of reference.

[198] Harvard educators George Homans and Lawrence J. Henderson were both heavily influenced by Pareto and passed on Pareto's teachings on social cyclical mobility to Talcott Parsons, another Harvard educator. Parsons wrote the American version of a textbook on sociology, *The Structure of Social Action*. George Homans, who was biologically related to both President Adamses, also wrote on human behavior with his *Social Behavior: Its Elementary Forms*. Henderson was an MD who taught chemistry and physiology and developed many contributions to both sciences as well as socioeconomics. He was incredibly brilliant, recognizing way back then the intimate association of these same ideas and taking Pareto's ideas to the next level. Henderson, like Zipf, "applied the concept of social systems to all disciplines that study the meanings communicated in interactions between two or more persons acting in roles or role-sets." *Wikipedia*, s.v. "Lawrence Joseph Henderson," last modified April 18, 2014, http://en.wikipedia.org/wiki/Lawrence_Joseph_Henderson. Places like Harvard are by no coincidence power-law kinds of places, where like attracts like.

two knew of each other, I do know they each had similarly evolving ideas of these repeating imbalanced occurrences,[199] and they both believed as I do that these ideas were disproportionately worth their value and should therefore be shared with the rest of the world.

In the next few sections, I therefore plan to highlight a few other examples of the principle of imbalance and show how those examples relate to Zipf's principle of least effort and our unconsciously estimated human efficiency. In particular I wish to highlight how these same principles can relate to our choices in geography, marriage, and educational achievements.

---

199 Zipf even makes reference to predicting the size of the elite class that Pareto refers to as a function of the square root of the population size (a number Rousseau also noted). George Kingsley Zipf, *Human Behavior and the Principle of Least Effort*, (1949; repr., Mansfield Center, CT: Martino Publishing, 2012), 452. For the United States, the square root of the population would be roughly 17,500 people, and some multiple of that would be perhaps four to five times that. Those would be the real shakers and movers, and I am not sure exactly what percentile that would be (about 0.005 percent), but it is not the top 1 percent (that would be three million people in America currently, or one million households roughly!)

"Where Shall We Live?"

Geography and Zipf's Principle of Least Effort

Some would philosophize that humans exist only because of where we fortuitously live within the geography of the greater universe.[200] Conditions had to be perfect and within a certain ideal range to allow human life to exist and flourish over enormous spans of time. Previously I thought you could live anywhere and make it work (humans are amazingly unique at reshaping our geography), and while there is some truth to that, conditions have to be initially somewhat favorable for life and hospitable for continued survival, as we know from the larger-scale universe frame of reference (FOR).[201]

Balanced with this idea of limited favorable geography is the reality of an exponentially expanding world population, which contributes to the need to understand these principles of efficiency. These principles are often the hidden reasons for our arguments both within society (as in the case of these disparities in income, education, and health) as well as on a larger level with wars and conflict.[202] We are all competing *and* cooperating to

---

[200] Others would even go so far as to say that there are infinite multiverses and that we live in a bubble that is one version of all the probabilities that can occur. Conditions are favorable in ours, which is why we exist. At some point in this book I show the incredible and fortuitous skewed outcomes that favor our existence on this naturally imbalanced planet.

[201] This is another nod to my title and another reason I chose the word *for*. More for less is all about relativity, and you need a frame of reference (FOR) for everything in life, including geography. Distances on a map are always relative to some reference. Similarly, how we compare the separation between ourselves sociologically is similar and thus geographically dependent.

[202] Thomas Malthus, *An Essay on the Principle of Population* (London: J. Johnson Publishers, 1798). Malthus's work later influenced Pareto (who references Malthus in his topics of exponential population growth) and Darwin as well as others. It can be a civil war or an international war—it is all the same relatively speaking.

obtain limited resources in a growing Malthusian[203] world that requires more efficiency. Wars and diseases[204] by no coincidence follow power-law distributions independent of time and geography. Wars and diseases simply represent the same fundamental human behavior[205] of society's soldiers, sailors, and doctors from the nursery rhyme.

The distribution of these outcomes (both desirable and undesirable) is therefore no different on a smaller scale when we examine where we *choose* to live within our earthly frame of reference, either in our home countries, our states, or within smaller communities. The skewed outcomes in measures like diseases otherwise involve the same elements, which become similarly imbalanced, although negatively.[206] This is particularly interesting to me since I spend my life treating diseases as a career choice. Disparities in the outcomes related to geography not surprisingly abound in the same way as other measured human outcomes (both good and bad), so you should clearly take a lot of time in choosing where you live, factoring the principle of imbalance into your choice.

Pareto first demonstrated that wealth and income both follow predictably imbalanced distributions—a few people are able to earn a lot and possess lots of wealth, and a lot of people earn little and possess little wealth. If you will recall, Pareto hypothesized that this was not unique to his own choice of geography (Italy), and so he examined income and tax information for other countries, showing the same mathematically skewed distribution everywhere. Pareto found the distribution was

---

203 Thomas Robert Malthus was a cleric, a scholar, and an avid student of population outcomes. He proposed that a balance of our world population was achieved through famine, disease, misery, and other "naturally" occurring checks and balances that had divine input as a way to teach virtuous behavior. He also suggested other restraints such as celibacy and postponement of marriage as ways to keep population growth in check, particularly in already poor areas. His ideas led to the creation of the national census in England, from which Pareto would be able to later abstract data.

204 I show the relationship of diseases within America in the appendix.

205 I show later the power-law distribution for wars, which applies to both civil wars (like in our own American history) and in world wars (which I show as an example in the appendix). The scale is the only difference.

206 I could argue geography is just a grouping mechanism of looking at human nature. It forms its own reference frame, as do the others I describe, and they all therefore probably measure the same things, just as different manifestations.

remarkably universal, in his words, "through any human society, in any age, or country."[207]

The spread of income and wealth, as Pareto theorized, is therefore predictably imbalanced *wherever* you live. Some might take this to mean that the imbalance is not unique to where you live, so you might as well live where you want! But it is not always so simple, as I learned the hard way. Sometimes hidden forces keep us grounded in certain unproductive geographies due to something known socially as *propinquity.* Other forces draw us to areas of efficiency[208] and productivity like a social law of geographic attraction. I wish to show you what to look for and how these efficiencies (or repeated inefficiencies) in our efforts relate to these imbalances within specific geographies.[209]

Hidden variables are indeed at work everywhere, and some geographic choices are just naturally more appealing than others—just like earth is more favorable and preferable[210] as an environment for life than Mars. In a more earthly frame of reference, Africa may seem unappealing to many Americans due to consistently unfavorable and sometime hostile geographic-related outcomes. Even though it may be achievable in theory to eventually inhabit other planetary geographies, would any of us currently choose to use disproportionately more effort to live on Mars knowing what we do about the harsh conditions found there? Or likewise would you choose to live in Africa, knowing it would take a lot more effort to meet your basic needs? Would you choose to live on the crater of a volcano, even though technically it is something you could achieve? Or in my case, should I remain in a state with repeatedly poor outcomes in regard to what I value with education and occupations and general happiness? While these examples may seem ludicrous, I assure you they

---

207 Pareto, *Cours d'économie*.

208 Similarly, the existing population could make efficient choices to reshape the imbalances locally if they knew this idea and made the appropriate inputs.

209 This is analogous to what Zipf called forces of unification (bringing things together) and forces of diversification (splitting them apart). They form a sort of dynamic equilibrium that is defined by these mathematical laws. Geography is just another example.

210 I would point out favor*able* and prefer*able* are both human abilities that follow Pareto's ability curve.

are not, and they typify the very nature of these skewed life outcomes in many related ways.[211]

Geographic factors, while perhaps not as emphasized in Pareto's principle of [income] imbalance because his principle is seemingly independent of location,[212] *are* important in the sense that the same distribution and patterns of human settlements within a population exist by these same principles of efficiency. People choose to distribute themselves geographically in the same imbalanced way as money (perhaps people learned to follow the money). Stated another way, we are as imbalanced in our distribution by geography as we are in our distribution of income. Either way you look at it, we are skewed—in Italy (Pareto), Germany (Zipf), America (Juran), or elsewhere! So there is no right or wrong choice by nation, although you might choose to pick one that is more efficient or less efficient than another, depending on the outcome you are after.[213]

Zipf, who was greatly interested in our human ability to communicate ideas with each other, analyzed much more than just words to show cultural repetition and our choices. To Zipf, the skewed nature of things like money and word use was more a statement about human behavioral efficiencies (efforts) and less about the money or words themselves. To him they all represented tools and users. So Zipf examined other systems like geography, marriage partners, newspaper media, music, and other variables that we generally value within popular culture. What Zipf discovered was that similarly skewed imbalances fit power-law distributions regardless of what

---

211 West Virginia is consistently in the bottom 20 percent of every measured positive outcome I could find. Similarly Africa is in the bottom measure of most valued societal preferences. Mars is an extreme example, but in a thousand years, we may not think so. The point is to identify the geographic factors in each area that favor our existence and our species growth.

212 In all actuality the degree of inequality does vary based on geography. Scandinavian countries as a rule enjoy less disparity for income.

213 For example, as I state elsewhere, Scandinavian countries boast less income disparity and greater social well-being (i.e., win-win outcomes) and happiness in general, which would include other variables like good health, free choice, quality of education, life expectancy, and so on. They are patterned to be less competitive and more cooperative, which is represented by their governing beliefs (e.g., more socialist). So if you desire more equality, go live in Denmark or Sweden or Norway (assuming they want you), or change your own government by your electoral process. But living in a nonsocialist country (e.g., America) will pose challenges in this regard, as you might imagine.

nation or time period you examined. Languages followed the same spread as money (perhaps "Money talks") and clustered in the same ways that people congregated in geographic locations. If money and people and other variables correlated with the same geographic patterns, then the rules are the same since the inputs are all human inputs (i.e., behavior).

What Zipf discovered regarding geographies was that people tend to settle and live in and around large metropolitan areas in predictable ways. For example, 20 percent of the larger cities within the United States currently account for the majority of the American population (roughly 80 percent),[214] and 80 percent of the geographies in Zipf's analysis (i.e., smaller towns like mine) accounted for a much smaller component of the population. My community and my state are great examples of this. Most of our state population clusters in a few larger cities,[215] and the majority of our numerous towns harbor few people. Our family would be way out on the geography curve described by Zipf in rural, hard-to-reach areas that represent the growing population minorities of America (although in the old days, that was the typical all-American community).

In other words, in today's America, *big cities are more popular*, and small cities are not.[216] And since growth is often exponentially proportional to existing population bases, the disparities only get worse with time, especially when you factor in other variables besides just birth rates and deaths. Income, education, jobs, entertainment, and so on tend to add to the layers of complexity and make cities grow even more disproportionately

---

[214] US demographics in fact show 82 percent of Americans live in cities or suburbs, as compared to the world urban rate of 52 percent. A few large metropolitan centers account for the majority of these population skews as well (New York City is the top of the social population pyramid). US Census Bureau, http:// http://www.census.gov/popest/data/index.html.

[215] Having said this, the largest city in West Virginia (Charleston) has roughly 50,000 people. Within this small state of 1.8 million people the (Zipf) rule still applies. US Census Bureau, Census 2010. There are 205 (more) towns with populations fewer than 5,000 (less), there are 13 cities with 5,000 to 10,000 people, and 14(few) cities with populations of more than 10,000 (more). These population numbers fit a classic Zipf power-law distribution with exponent (power) of −1. In addition, roughly 20 percent of our cities account for 70 percent of our population.

[216] Zipf appropriately references Robert Gibrat, another discoverer of this fact. The relation of city size to its proportional growth (log-normal when graphed) is sometimes referred to as Gibrat's law. It is not quite a power law as Zipf showed in his data, but the idea is similar.

based on what we are attracted to. Big cities (like New York City) get bigger, and small communities get smaller. It is somewhat analogous to another observation: "The rich get richer, and the poor get poorer," which is based on similar mechanisms of growth.

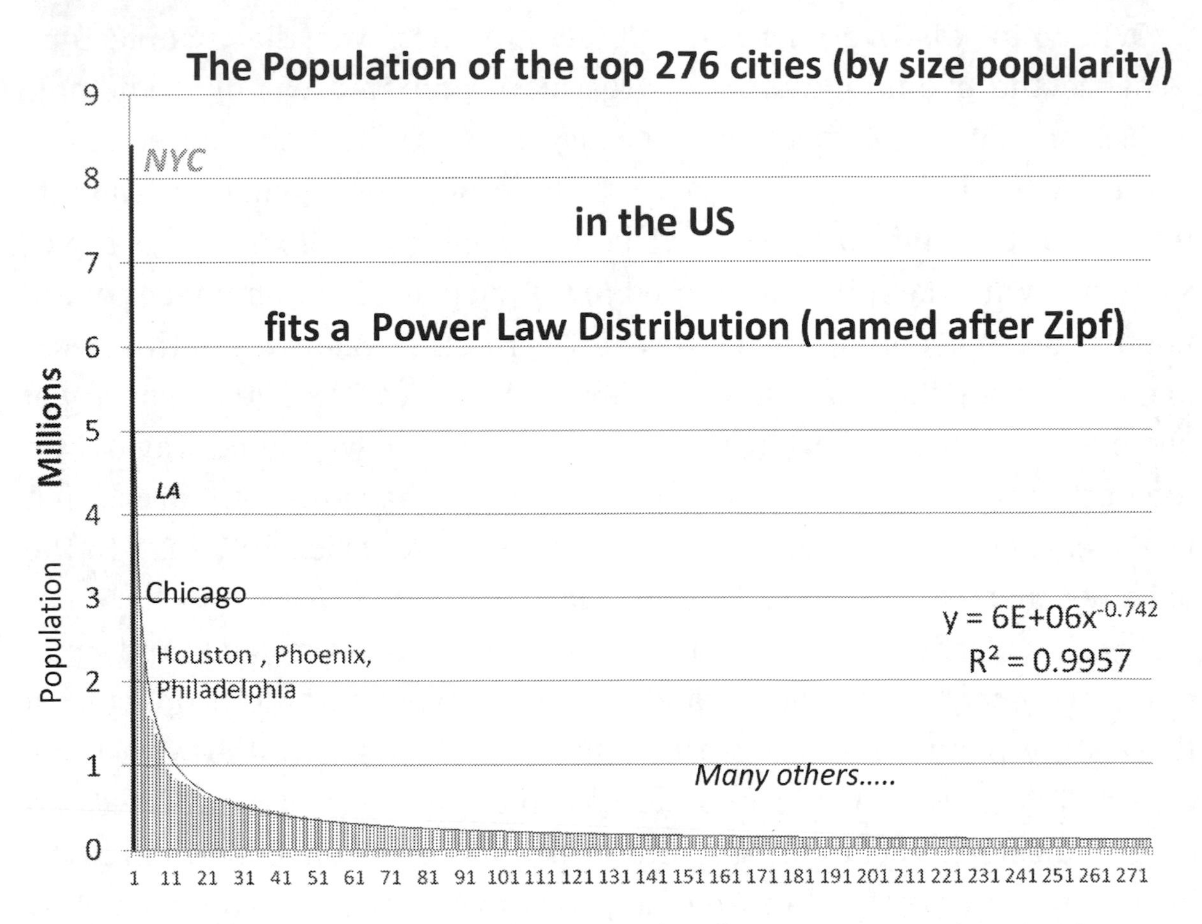

This graphic shows the skewed distribution of people by popular cities in America. This is an example of a Zipfian power-law distribution for geography.

Certain geographies are clearly more populated (more popular by choice) than others. It may be due to something as intuitively simple as the relative abundance of drinking water, food, clean air, shelter, and the more basic needs we physiologically and biologically require as suggested (and coincidentally ranked) by psychologist Abraham Maslow.[217] Or it may be something more complex like the type of symphony or sports stadium the city hosts or the type of college or industry it serves. Either way, the outcomes follow the same repeatable patterns of skewed human choice and consequence that Zipf noted over half a century ago.

For the United States currently, a plot of the simple population density of America's largest cities by rank, as I showed, reveals a typical long-tail, skewed distribution that is now a familiar distribution to you as a power thinker. This phenomenon nicely demonstrates the principle of least effort in relation to our geographic choices. *A few American cities account for a disproportionate amount of our people.* It is certainly easier to live where everything can be found relatively close by than to live somewhere that jobs, food, resources, and so on, are not as readily accessible.

One could hypothesize therefore that if humans are the input and are disproportionately located in the United States, the other measures we are interested in should be similarly distributed (like education, money, health, happiness, and so on). In other words, people choose their behavior, and the outcome is simply the result of their choices—in this case the population distribution of people around a few large cities (with necessary resources) is the most efficient model for a growing society. This same model can be seen to apply everywhere else in the world and not just for the United States.[218]

In the case of geographic preferences, a few of the largest American cities account for most of the population of the United States with obvious disparities noted, and the relationship nicely fits a power law. New York City is notably the highest in population density. It is by no coincidence that incomes are also notably better on average there, as well as the concentration

---

[217] Maslow described a hierarchal system of human needs in psychology in the 1940s and 1950s. His five-stage model began with the basic physiological needs (food, water, sleep, etc.) and progressed to the final level of self-actualization (realizing personal potential and self-fulfillment). In his model, only one person in one hundred (the 1 percent) reached the final level due to the disproportional rewards society places on motivations. A. Maslow, "A Theory of Human Motivation," *Psychological Review* 50 (1943): 370–96, http://psychclassics.yorku.ca/Maslow/motivation.htm.

[218] Part of this relates to the mechanics of the exponential growth of an existing population. But the key is that the resources and efficiencies must support the growth; otherwise this phenomenon is not seen (like where I live).

of wealthy (the rich man), including billionaires (for the United States as well as worldwide[219]) and many other successes that my kids would admire (athletes, superstars, models, etc.). If I aspired to become a billionaire or indeed just very wealthy or wished to have a higher likelihood of a certain job, a better career, or a certain lifestyle that may be more unique than the *average*, I would logically consider moving somewhere like New York City to improve my odds of exposure. This may explain my sister's choice of New York City for her home; she has learned to be more efficient in her choices than I have![220]

Examining this same idea on a larger reference scale, at the country level for example with the G-20 (a group of industrialized nations, including Europe and the United States among others), we see similar geographic distributions of wealth and population are found throughout the world—20 percent of the world's countries control 80 percent of the entire world's wealth; similarly, a few countries, like China and India, account for most of the world's population. It is by no coincidence that these numbers fit descriptive laws much like those Zipf and Pareto discovered over a century ago, in different cultures and in different countries. People choose to be (or smartly go) where the action can be more effortless and where they can be more efficient in acquiring jobs, food, health, and everything else they value.

I could argue that most of the variables I discuss (money, education, health outcomes, religion, etc.) are measures of the same thing—our basic needs and other preferences matched with our efficiency in our choices. That is what George Zipf argued in the 1920s–1940s with the principle of least effort in regard to the mechanism of human behavior. Whether it is our choice of words or our choice of geography, the outcomes are the same functions of human behavioral efficiencies. Some people learn to be more efficient, and others struggle their entire existence.

---

[219] The United States has a disproportionate number of the billionaires in the world, and New York City has the most billionaires per city by far in the United States (and yes, I plotted it, and it fits a power-law distribution as well).

[220] It must be the influence of Harvard in my sister's PhD educational achievement that taught her this concept. The one office building where her husband works (for billionaire and former mayor Bloomberg) has as many employees as my entire small-town community!

While this may seem intuitively obvious to you, it was not to me at first. I *assumed* in modern-day America that most of the basic geography-related issues had been (or could be) resolved. In a modern industrialized nation like America, it is hard to imagine that such great inequalities could exist between our smaller communities for things like health outcomes, in our counties for things like educational-attainment results, and within our larger states or regions for things like jobs and incomes.[221]

Your choice of geography can as profoundly affect your occupational success, your marital happiness, your health outcomes, and your children's success as many other inputs I have discovered. I *chose* to live in a small city (way out on the Zipf power-law curve long tail where the population is 3,800), initially thinking I could avoid some really bad inputs that correlated with a bigger city, such as crime. I was wrong in my simplistic and assumptive thinking because, as it turns out, the results of just about all these outcomes are not as much city-size dependent as tied to other more universal concepts like income disparity, lack of money and opportunity, or more simply principles of human (imbalanced) nature, which are culturally resistant to change everywhere. What most likely contribute to the observed effects are the perceived differences in what we desire relative to our own reference frames. So for my provincial-minded community it would be the local perception of disparity relative to our own frame of reference rather than when we look at a province in Africa (e.g., Rwanda) or some other poor country.

However, I believe the size of your city and, more importantly, what it places value on as a miniculture does matter to a large degree. Although Zipf did not say this directly, I suspect he considered this idea as well. Certain critical thresholds are needed for population growth and for these other outcomes to naturally follow. Certain ideas need to be valued geographically, such as higher education. Just like people need some minimal income for life, a city needs a minimal infrastructure for continued growth—and allocation

[221] This is an important point, though. We can all learn equally what to do, but the starting point is so low for certain geographies that they never really catch up. They consistently remain the bad performers since growth is often proportional to the existing population. The only way for them to catch up is to be more efficient in some way.

of resources to unpopulated areas from other states and from the federal government are unpopular choices to make in today's world.[222] Cities or states that become stagnant or decay in population[223] are often declining in these other outcomes as well, and with rhyme and for a reason. Size does matter in this sense, at least to a minimum degree. What that critical threshold is, I am not certain, but a pattern of growth from a historical perspective is a good indication of the dynamics and should be considered when making these choices. Cities and states with positive growth are more likely to be favorable than cities or states with negative growth, or decay. Which would you prefer from the practical sense—a decaying geography[224] or a flourishing one?

In a similar way, people of similar socioeconomic backgrounds and cultural beliefs tend to cluster around each other geographically, and the other outcomes seem to naturally follow. The people in your geography can and do affect the nature of your nurture, whether you are cognizant of this or not. People who are unmotivated tend to gravitate to others with low motivation. People who are achievers tend to attract or move toward those with similar-minded abilities. People with poor health habits tend to congregate together when they choose to repeatedly make unhealthy decisions. These *laws of attraction* are society's geographic version of gravity, and they invisibly guide our interactions with others everywhere.

As a general rule, 50 percent of Americans live within fifty or so miles of their birthplace, a social concept known in psychology as

---

[222] This was one of the reasons my state broke away from Virginia during the Civil War era. West Virginia did not feel its share of the booty was fair, just as people feel now. This discontent with the ability of Virginia to redistribute its wealth to the more needy and less fortunate was probably more integral to the separation of the two states than the issues of slavery. Money is usually at the root of more issues of that nature, with other factors contributing less.

[223] The population of my already sparsely populated geography is projected to decay in all age groups from eighteen to sixty-five over the next twenty to thirty years. The only projected age group that will increase is those over sixty-five, mainly due to the cheaper-than-average lifestyle in West Virginia in big-ticket items like housing (lowest average cost for a home of all fifty states).

[224] This is what is meant by ghost towns and dying communities. My community, for example, has a 1 percent growth rate. The state has a 0.1 percent growth rate for this year, and that rate will be negative in the next ten to twenty years. In comparison, the United States as a whole has a growth rate of 1.7 percent currently. This is a problem. US Census Bureau, http://quickfacts.census.gov/qfd/states/00000.html.

(geographic) *propinquity.*[225] And while you have probably never heard the word, propinquity greatly affects your opinion of others, of society, and of the greater world. We are likewise socially close to people that we work with most often (occupational propinquity), those we attend school with (educational propinquity), and those with similar educational attainment, religious values, and so on. All this can even affect who we one day may choose to marry.[226] Unfortunately, when things are not favorable (like inhabiting a highly drug-infested area, a gang-ridden community, or a highly uneducated geography), some of these unfavorable outcomes continue to be propagated and perpetuated by like attracting like, unless more-efficient changes are sought and implemented.

To my simplistic way of thinking, I sometimes ask myself, "Why not just move if conditions are so bad?" If most people around you don't want to change the way things are locally (or cannot for reasons of limited resources), then sometimes you need to change your geographic choice for your own health or happiness. However, you must realize that propinquity is a powerful force that can keep people grounded, even when it makes no sense to others (and like gravity, it is often an invisible force). As an extreme example, when natural catastrophes (e.g., hurricanes) are predicted to occur, people often choose to stay at home and ride out the storms despite mandatory evacuations and the risk to health and life. This reluctance to leave is due to these same strong social connections with their only familiar geography, as well as some assumed calculated risk that they will be unharmed.[227] I also recognize that some people simply do not have the luxury of moving, and this factors into the equation, but often geography is merely a matter of choice. Hopefully you get the idea. Geography as a choice can have many profound hidden implications, and I discuss this again in

225 I have also seen it referred to as viscosity. The US Census Bureau shows that on average 59 percent of American people live in the same state in which they were born.

226 Donald M. Marvin wrote his PhD dissertation on this very subject, "Occupational Propinquity as a Factor in Marriage."

227 Similarly, geography becomes a unifying force when it comes to our nation, especially when we feel threatened or unfairly treated.

the next section with how it relates to your choice in marriage, and I look at propinquity as a law of social attraction.

As a personal example of how geographic propinquity relates to American educational values, the educational efforts (or relative lack thereof) in the area where I chose to live for the past two decades have negatively and disproportionately affected my children and their education. I chose to live and practice in a state with the *lowest* levels of educational attainment of all fifty states.[228] I chose to live there in order to stay close to my family of origin,[229] which once more relates with this idea of propinquity as well. And while it is not simply that we are mostly stupid in West Virginia, it does concern me that we are unable to learn to make the appropriate changes needed as a state over time. This implies certain inefficiency invisibly linked to our local geography. Let me highlight this in a few ways that emphasizes West Virginia's own geographically related values ("efforts" as Zipf would suggest and "abilities" according to Pareto).

In order to see Zipf's principle of least effort in West Virginia in action and to see these imbalances in relation to other states, I chose to examine (1) the percentage of the state population that achieves higher learning through college attainment and (2) the time it takes to achieve graduation in a typical West Virginia university. Both are simple measures of cultural efforts and efficiency.

Based on twenty-year data from 1990–2009 from the US Census Bureau, West Virginia has consistently achieved the lowest levels of education within America. And while we tend to blame these consistently poor outcomes on farming and the emphasis on other manual labor jobs (e.g., coal mines) that don't require higher education, these jobs are not currently the most common in our state. And so our inabilities perhaps represent old learned ideas from previous generations when these former occupations were popular. Disproportionately more people currently work in minimum-wage

[228] I am ashamed to say I did not know this when my family moved to the area. I assumed we would be able to shield our children from poor education with private schools, but private schools are also underperformers in our geography.

[229] We wanted to remain close to our aging parents to be available in their declining health.

occupations[230] in the food and related service industry (food preparation is one of the lowest-paying occupations, as I show elsewhere), and neither farming nor coal mining are as popular as they were previously. However, the cultural mind-set of not valuing higher education remains the same.

How we have "Learned" the Principle of Least Effort

Year Examined	1990	2000	2009
HS grad(WV)	*66%*	*75.2%*	*82.8%*
(US average)	**(75.2%)**	**(80.4%)**	**(85.3%)**
(rank)	48/50	47/50	44/50
Bachelor's(WV)	*12.3%*	*14.8%*	*17.3%*
(US average)	**(20.3%)**	**(24.4%)**	**(27.9%)**
(rank)	50/50	50/50	50/50
Advanced Deg(WV)	*4.8%*	*5.9%*	*6.7%*
(US average)	**(7.2%)**	**(8.9%)**	**(10.3%)**
(rank)	48/50	47/50	49/50
Income(WV)	*$35,305*	*$38,858*	*$38,100*
(US average)	**($51,028)**	**($54,951)**	**($51,100)**
(rank)	49/50	50/50	49/50
Disability(work)	*12.6%*	*13.2%*	*9.5%*
(US average)	**(8.1%)**	**(11.9%)**	**(4.6%)**
(rank)	#1 highest	#1 highest*	#1 highest

[230] Our state is in the top fifteen states for minimum-wage workers as a percentage of our work force. US Bureau of Labor Statistics 2012 data shows 5.7 percent in West Virginia versus the national average of 4.7 percent. This is not a good statistic unless your goal is to work a minimum-wage job. Most of these minimum-wage workers are skewed toward the youth as you might imagine. The challenge I see is how to get people who are disabled from work back into the workforce if their only real option (with lack of education) is a minimum wage or low-wage job (the median per capita income in West Virginia, at $22,000, was forty-ninth in the United States from a 2008–2012 American Community Survey five-year estimates). Minimum wage at 2,080 hours a year comes out to $15,000 per year. Disability income (SSDI without working) comes out to $12,000–$13,000 per year for an eligible couple and even less for the individual, so given the choice of not working and getting nearly the same as working a minimum-wage job with a lot more effort seems to be a no-brainer, and most people have figured that one out in our state.

I graphed here a few vital statistics that highlight some of the geography-related educational and income disparity seen in West Virginia with work-related disability shown as well. The results in *gray* are from West Virginia (and shown for the various decades), and the **black** serves as reference from the average US data from all fifty states during the same time periods for high school attainment, bachelor's degrees or above, advanced degrees, median income, and disability.[231]

I show this graphic to illustrate several points. First, West Virginia as a state achieves the lowest levels in the nation for many good measures yet has the highest levels of disability and welfare.[232] This has become a *learned* part of our local culture in the sense that we have come to expect the federal and state government to take care of us as a population. We expect more, but we contribute less than everyone else to achieve it, based on low population numbers, low educational attainment, low-income earnings, and thus a poor tax base. Yet we have higher-than-normal expectations for the rest of the country to take care of us. This is in essence the same idea of income redistribution, wanting to take money from the wealthier states to support the poorer states that consistently choose inefficiently in their outcomes for many measures!

Secondly, for the few youth in our typical community that do choose to matriculate to colleges within West Virginia (out of state is prohibitively expensive if you are poor), they struggle with the time constraints of college

[231] Income data comes from http://nces.ed.gov/programs/digest/d11/tables/dt11_025.asp data for income based on 2010 dollars; education stats are from the table from the US Census Bureau, Table 233: Educational Attainment by State: 1990 to 2009, http://www.census.gov/compendia/statab/2012/tables/12s0233.pdf; work disability data is from http://www.cdc.gov/mmwr//preview/mmwrhtml/00021981.htm#00000172.htm. The disability rates I selected were for work-related disability and not sensory disabilities or physical or mental handicaps. In 2000, which I did not list above, the data was mixed. Judith Waldrop and Sharon M. Stern, *Disability Status: 2000* (US Census Bureau, March 2003), 7. The total overall disability rate was highest in West Virginia out of all states within America at 24.4 percent; for work disability we were 13.2 percent, which was in the top ten worst. A lot of the disability comes from the median age of our population being older and age-related chronic diseases. Also a hard physical lifestyle from mining may certainly contribute as well, as do poor health-related choices like tobacco and obesity. An epidemiologist could write an entire book about the *state of our state* as a symptom of the greater problem within America.

[232] West Virginia Medicaid rates are the highest in the nation, at 28 percent of the state's population.

with poor efficiencies and delayed graduation. As an example of this, only 36 percent of college-goers at West Virginia University graduate within four years, and only 57 percent graduate within six years, which compares rather unfavorably to surrounding state and national school averages.[233] Sadly, some of our youth are actually proud of this statistic.[234]

The biggest challenge I therefore see within my local culture is a lack of desire or knowledge of how to efficiently change these outcomes. This is in fact creating a projected exodus in youth from West Virginia that will of course only further propagate the state's problems over time rather than improve them. Joseph M. Juran, another tinker whom I will discuss further in part 3, hypothesized that cultures in general are resistant to change, as not changing involves the least amount of effort. This resistance to change is a truism wherever your choice of geography, although some locations are clearly more resistant than others.

So what do I conclude from this geographic disparity in important variables like educational attainment, work and careers, and earnings? I want my children to be surrounded by people that are more like them than not. I want college to be a learned expectation and not a geographic or cultural exception. I want them to learn to be self-reliant and not expect the government to take care of them for free. I want my kids to gravitate to those who help bring them up (in things like motivation and educational attainment and achievement) and not bring them down. This is a tough concept to learn early in life, much like gravity is a tough law to learn, but

---

[233] Two hours away but across state lines, the University of Virginia recently ranked number one in public institutions in four-year graduation rates at 87 percent, with six-year graduation rates at 94 percent (stats from www. kiplinger.com). University of North Carolina also did well with 89 percent in a recent 2013 report. West Virginia University was one of the lowest in the country as a flagship school with 32 percent four-year graduation rates and 54 percent six-year rates. West Virginia University was also ranked 170 for national universities in a recent ranking for 2013. More school information can be found on US News and World Report rankings of colleges from http://colleges.usnews.rankingsandreviews.com/best-colleges/rankings/national-universities/data. It is amazing how much geographic disparity can exist in our ability to graduate our college kids. Should we as a society redistribute the quality of our education to change this? Or take away money and great teachers from other schools to make this happen more efficiently in states like mine? If you are on the less educated side, you may say yes; if you are on the more educated side, no. It is mostly about perspective.

[234] Our state university has a "proud" reputation of being a party school among the youth in our region.

the basic concept is important. Break out of the current mold if you don't feel it fits with your needs. If the local people do not want to make the appropriate changes or the state or federal government cannot help them make the changes, your only option is to change what you are able to do relative to others and help educate them however you can. Similarly, people that bring you down in your standards or that perpetuate and encourage negative behavior are not good choices, so separate yourself from these people and, if necessary, from those geographies.

I want my kids, like Daisy, to learn a new way of critically thinking about their lives and examining their geographic choices by learning to ask the right questions. The choice does not have to be between the *big house* or the *barn*,[235] but it needs to balance the issues of geography-related disparities that link to their career opportunities, desired means of income, education, marriage, religion and core values, and general well-being. I want my children to find what it is that attracts them and then lead themselves to it more often by making efficient choices that are disproportionately rewarding for them.

The current gestalt among power-thinking experts is that the new generation (my kids) will have on average *six different jobs* over their careers.[236] This will necessitate some willingness to move for the jobs that match your unique abilities, as opposed to staying in place and hoping the government will help you with your outcomes. It will also necessitate the same skills that allow more flexibility to achieve this, including an enriched education. Zipf would use the analogy of *tools seeking jobs* (e.g., a carpenter looking for work wherever he can use his hammer), but another way to visualize this is *jobs seeking tools* (e.g., someone recruiting you for your

[235] There are not too many big houses in West Virginia. The state has the lowest median value for homes at $94,500 (a lot of little houses and barns) in 2012, which on the positive side results in the highest percentage of home ownership in the country (number one at 75.8 percent). http://www.census.gov/hhes/www/housing/census/historic/owner.html; and http://quickfacts.census.gov/qfd/states/54000.html.

[236] Levine and Dean, "Education for Life in an Evolving Information Economy," in *Generation on a Tightrope*. Six careers is the new average, and one or more of those jobs may not exist yet due to the rapid changes in technology that we will continue to see grow, so the young generation needs to be dynamic. The best way to achieve this is through an enriched major in college, practical minors, and internships, according to Arthur Levine.

own unique skills that then allows you to select where you live). Ideally, you may eventually enjoy a career like mine that is mostly independent of your geography (i.e., one that is needed everywhere and always, like a doctor, lawyer, merchant, or chief) or totally independent of geography (e.g., a writer, artist, musician, or Internet business). The ideal job will allow flexibility based on what you desire rather than what the geography dictates or will support through supply-and-demand economics.

Finding one and only one geography that fits that entire bill may be difficult in today's ever-changing world, and the reality is that the geography may change (and more than likely will) as time goes on as a function of many factors. Sometimes it is more efficient for you to unlearn some of your internally programmed propinquity for geography. What may work for the first twenty years may not be ideal for the last twenty, as I have discovered. You need to be *dynamic*[237] and learn to use power thinking to your advantage, whether that is here in America or in emerging nations (e.g., BRIC countries) that are following similar rules of efficient imbalanced growth. But obtaining the best skill set you can that allows you this freedom is most of the solution, and getting this early is helpful.

The top few largest and fastest-growing populations in the world currently are *BRIC nations,* and they cannot help but be major players in the world arena over the next twenty to forty years. If you collectively examine their combined population power, they were responsible for 2.81 billion of the estimated 6.7 billion (from 2008 statistics) people on our planet, and they will represent 3.2 billion out of an estimated 8 billion by 2050. Just four countries represent a very imbalanced proportion of the world population, but this skew is still predictably imbalanced according to these same rules of geographic distribution (India and not China is predicted to be number one by 2050 due to its higher power of growth).[238]

In America, we have a projected population growth rate about 1–2 percent over my children's generation. I have seen in my own lifetime the

[237] The word *dynamic* has Greek origins and means "power." I am certain this is no coincidence.

[238] I should point out these issues factored into my wife's and my ability to adopt two of our four children in societies that previously had insufficient resources to tend to their population but have since improved (Russia and China).

average American life expectancy increase by about eleven years due to improvements in medicine, and if the trend in population growth and longevity continues, there will be an increasing need for similar future efficiencies in the way we live and choose to behave. If the trend in life expectancy continues as projected for the next twenty to forty years, living to ninety would not be unreasonable for the young generation. These changes in population growth and longevity require us as Americans to adapt in the following ways:

1. The young generation (my Daisys) needs to plan to work longer and/or save more efficiently to support a longer retirement period. If this is the case, youth should make sure they enjoy what they do and where they are.
2. The American government needs to account for this longer retirement period in its planning if it intends to continue to support the welfare system (with programs like Social Security), and it will need to begin to allocate resources more efficiently if it is to remain solvent.
3. The dependency rate—the relative percentage of people dependent on others (this generally is those younger than fifteen and older than sixty-five)—will grow as my generation ages. This number is expected to grow from a historically stable 30 percent to nearly 41 percent by 2040! This could affect my kids in terms of their need to care for us (their parents) if we become dependents, as many older people in our country currently are. (Don't worry, kids; you are not my retirement plan as I have always teased you. I have planned much better than that.)
4. The jobs and careers our children desire may in fact be elsewhere geographically (in areas of population and job growth rather than in areas of stagnation or decay), but we can help educate people within our geography to learn these concepts and hopefully input needed change.

From the principle of imbalance and the principle of least effort, I want my kids to realize that these hidden social forces invisibly shape our outcomes. If we are aware of these geographic factors (like propinquity), we can make the necessary changes to shape the outcomes we desire, whether that is something as simple as where we choose to live or which peers we choose to associate with.

Within the larger frame of reference, if we wish to avoid misery and famine and avert a Malthusian catastrophe as a nation, we need to learn to be more efficient in our choices and with our values. If we wish to change America, we need to learn these ideas now. There is no time like the present to start applying these rules to our lives.

One of my family's mottoes is "We can do hard things." Yes, we do mostly prefer easy, and as Zipf argued, using the least amount of effort is a natural tendency for all humans and not just for certain geographies or certain families. But in order to achieve the desired quality of many things that are valued in society and within our family, my wife, kids, and I recognize that we must put forth significant effort. Ideally the effort needs to be more than some reference minimum level and certainly not the *least,* as exemplified in our current geography.

In the next chapter I wish to highlight the principle of least effort further by how it relates to other relationships, specifically marriage and the choice of a life partner.

"Who Shall It Be—Who Will Marry Me?"

Marriage and the Principle of Least Effort

As inferred by "Tinker Tailor," our choice of spouse will make a big difference in our happiness and in our well-being—in more ways than just for income.

Your choice in a life partner will help determine not only where you live, the kind of house you inhabit, the clothes you wear, what you drive, and where you travel, but more relevantly whether you go to church or not (and thus the role of religion in your life), how many kids you have, how you care for your family of reference (e.g., *grandmother* from the rhyme), and what other values you choose to model as a family.

The implied lifestyle outcomes of marital harmony, family cohesion, personal security, spiritual development, and general well-being all follow this important choice. And although a spouse per se does not make a person happy or direct a partner's personal development, I cannot help but argue the point that spouses are hugely important as a cofactor in every outcome people are able to achieve in their adult lives. The power-law rules once more come into play in terms of a vital few choices. This is an area where the principle of least effort must be well understood as an invisible rule.

Since Daisy seemed so fixated on marriage in "Tinker Tailor," I wanted as a father to also include this section on marriage to further link these ideas for my own Daisys (and my son as well) about this disproportionately major life choice. While perhaps marriage and what defines the family is different now than it was in the time period of Daisy, the time is approaching soon for my children (my oldest is almost twenty) to contemplate these issues, and marriage definitely needs to be planned and if possible analyzed with some efficiency in mind.

Referring back to what Pareto suggested in regard to how we rank everything within our culture, and considering the choices that my kids soon will face, marriage should be one of the most important choices we hold high in rank. In fact, in the rhyme it is the number one question that Daisy asks, underscoring its relevance in rank. Since I am also naturally biased in favor of teaching the value of educational attainment due to my personal experiences, it makes sense to consider linking the concepts of education and marriage as one. For my three girls, I colloquially refer to this as getting their "MRS degree," meaning carefully finding a future husband.[239]

For anyone contemplating marriage, it is as vital of an input as any other choice you will make in your lifetime. Whom you choose to marry relates to the idea of propinquity I discussed previously, and once more the imbalanced outcomes relating to marriages follow these same power-law principles I've been discussing.

Remember this (you will see it again!): a few critical variables that you choose as inputs account for the majority of your life results. In life, with happiness as the desired outcome, marriage[240] is one of these critically nonlinear inputs that matters a lot, and disproportionately so! Your marriage, if it is good, can mean many positive rewards, and if bad it can disproportionately affect many other outcomes as well.

Marriage as a modern concept has evolved in many ways since my youth. Marriage within America is truly a dynamic concept. In the 1960s when I was born, between 70–80 percent of adults were married. By 2008, this number had fallen to just above 50 percent.[241] In many ways, the older concept of marriage is becoming obsolete, as seen in the opinions of the younger generation, 40 percent of whom say it is truly an obsolete idea. People now get married later, if at all. They have children out of wedlock eight times more often (41 percent now versus 5 percent in the 1960s). And

---

[239] I don't mean to exclude men here, but the concept originated historically with women due to the inherent imbalances in our male-female roles in society, which have of course changed. I could equally suggest the MR degree here as well.

[240] I could equally substitute the word *religion* or several others you might consider important in rank.

[241] Pew Research Social and Demographic Trends, www.pewsocialtrends.org; Pew Research Center, November 2010, pewresearch.org.

the modern family is no longer simply a mother and a father with a few kids in the same household.

This is in fact part of the problem with how our values have devolved in ways that are "popular" and that do not necessarily translate into good or healthy outcomes. In a way that is similar to income and career choices, what society as a whole achieves by popular choice is not often desirable. As an example, divorce and separation rates have tripled since I was a child,[242] and same-sex marriages are a relatively new and growing occurrence, both increasingly popular by outcome measures.

Whom we marry and when we marry are certainly two of the most significant choices we will make as adults in our lifetime. Our choice of a partner for life will be certainly one of the single largest inputs that will shape our lifetime result of happiness and success, and this choice by nature defines our offspring (i.e., our Daisys). It should therefore be no surprise that marriage is a *power-law phenomenon*. More importantly marriage will by nature involve the principle of least effort, and this is where awareness of these principles can help guide you appropriately.

Joseph Juran, a modern thinker who developed a common application of the Pareto principle of imbalance as it relates to our values within society (and whom I describe in part 3), would have agreed that marriage and the implied family unit was one of the most vital principles in his life. Although he was a workaholic because he truly loved his work, I can ascertain from my readings that he was a true family man as well, and he valued his one and only marriage. As appropriate homage to him (he recently passed at 103 years of age), I want to note his successful marriage. He was married at age twenty-two and remained happily married for eighty-one years[243] to the same loving woman! He similarly enjoyed, as I do currently, four children. I believe marriage and the value of family were a huge part of his

---

[242] Pew Research Center, Social and Demographic Trends, *The Decline of Marriage and Rise of New Families*, November 18, 2010, http://www.pewsocialtrends.org/2010/11/18/the-decline-of-marriage-and-rise-of-new-families/2/#ii-overview.

[243] He developed Pareto's idea into something known as the 80–20 principle (or the 20-80 principle) in which a few variables (accounting for roughly 20 percent of the total) are responsible for the majority of the results we seek (roughly 80 percent). I could say here that his vital choice in his one wife accounted for more than 80 percent of his years of well-being through marriage.

measured well-being.[244] He learned something *vital*, and we could all learn from him.

It is my hope that you will also have only one (efficient) choice in a life partner as Juran did (and as I do) and that this partner will contribute as significantly to your happiness. If you take time and effort to explore the relationship and use quality management principles (i.e., power thinking), I think you will be happy with the vital choice you make in this regard as well. In this instance *more* (happy) outcomes will come from *less* (fewer) marriages and divorces.[245]

Men and women should both consider the serious nature of their choices regarding their future spouses. This choice should involve as much effort as any other significant life decision and not be made on an impulse or with haste. I realize love is not always analyzable, but the implication here is that your choices and actions can be. Once more, we must be cognizant of the effects of propinquity and how these other power-law phenomena subtly weave their magic through a few vital choices and their disproportional outcomes.

For my own Daisys, for example, I would suggest they consider delaying this choice of marriage until they have graduated from college. Why? In part because of physiological changes that come after age twenty to twenty-five, when the brain fully matures. Secondly, marriage rates are now more strongly linked to education than they have been in the past, with college graduates (64 percent stay married) much more likely to be and stay married than those who have never attended college (48 percent).[246] So getting an MRS degree may not be as unusual a concept as you may have been patterned to think! It can be approached with the same idea of

---

[244] I read John Butman's biography of Juran, entitled *Juran: A Lifetime of Influence* (New York: Wiley and Sons, 1997), as well as numerous other references on his life. Interestingly Juran had a son who was troubled as a youth, and the implied cause was frequent geographic relocation and having a workaholic father. I find the similarities to my own life remarkably coincidental, although I can proudly say I have never been absent in my children's lives. I enjoy a career that disproportionately allows me to be home 99 percent of weekdays by six o'clock, and weekends are protected family time as well. I am, however, a workaholic while at work.

[245] Pareto had two wives, and while this is still a vital *few*, in this case less is (more) better.

[246] Pew Research Center, *The Decline of Marriage and Rise of New Families*.

efficiency[247] as your education or any of these other ideas I have presented to you.

Unlike our parents, whom we cannot pick but who heavily affect our outcomes as I have shown, a spouse is clearly one variable we can control in our efforts and choices. *You can pick your spouse.*[248] And, girls, carefully (i.e., with significant effort) pick you should!

I do not mean that we make little effort in picking a mate, but the principles are invisibly at work. What I mean is that Zipf's (efficiency) principle of least effort applies, much as we have seen in other areas of human social behavior, such as where we choose to live. Like our choice in cities versus rural areas, we need to pick our marital partners efficiently and get it right the first time, even if that means delaying it until the time seems right! We sometimes don't take enough time and effort to examine the person and thus make the appropriate choices. If the person is right for you, they will wait. In other words, we tend to act impulsively at times, which can lead to a less-than-ideal outcome of pregnancy, divorce, or poverty—or all three in the case of my own family frame of reference.[249]

One of the original studies that influenced Zipf's work was a study of five thousand couples that applied for marriage licenses within a small area of Philadelphia in 1931 (during Zipf's time period census data for larger areas was harder to come by than now). For all distances within one to twenty blocks, there was a power-law distribution (with power of −0.85)[250] relating

---

247 I would also point out that the brain is not fully developed physiologically until twenty to twenty-five, so to make a choice like marriage and expect it to be long-lasting while your brain and ideas and values are still evolving makes little sense, especially in a culture that places less value on maintaining a marriage. Wait until your brain is fully myelinated and developed.

248 There is an important concept here as well. You cannot hope to pick someone that you can change, which is where a lot of people make an error in assumption in marriage. It is better to find someone you are happy with and accept them the way they are, rather than thinking you will eventually be able to change them in an important way.

249 My mother was seventeen, acted impulsively, got pregnant, and attempted a first marriage. She would birth me, get divorced, and become poor all in a short span as a result of a few bad choices. What was unique about her, though, was she learned and adapted, made appropriate changes, and later achieved what she desired through efficient choices of higher education, a second marriage, and other family core values.

250 Zipf, *Human Behavior and the Principle of Least Effort*, 406–7. He used data originally described by J. H. S. Bossard, but Zipf later examined it to show a log-log linear relationship that is characteristic

marriage to the idea of geographic proximity. In this particular study, for example, most of the marriage applications (70 percent) were within a few (five blocks or 30 percent of the total number examined) blocks of each other. In other words, people congregating together due to their geographic residences tended to marry each other.[251] This is another example of the principle of imbalance, and in this case relates to the law of attraction and residential propinquity.

Another scholar, Dr. Donald Marvin, similarly studied a comparable population in Philadelphia and noted a propensity for occupations to heavily factor into women's choices in marriage, particularly as more women began to enter the job force during the last century. In his PhD dissertation defense, Marvin noted that women (even one hundred years ago) tended to gravitate toward men from similar colleges (educational propinquity) and to men from similar occupations. Doctors tended to marry other doctors, teachers other teachers, and so on.[252] This latter concept is known as occupational propinquity.

One thing to note here is that there is likewise a tendency of people to marry persons of similar race, religion, and national origin. These factors probably relate to geography once again, especially in the 1930s for Zipf (and certainly earlier as Marvin showed). But we still see similar clustering of marriages today in ethnic groups, neighborhoods, high school sweethearts, colleges, and workplaces. It seems logical that we are more likely to meet and eventually marry someone close to where we live and work and socialize. So it follows that geographic clustering would correlate with marriage, leading us back to the idea of propinquity.[253] The point here

---

of power laws.

[251] This is an example of residential propinquity. It is probably not as close as blocks in the modern world, due to increasing abilities to travel farther, but the concept still is valid.

[252] Donald M. Marvin, "Occupational Propinquity as a Factor in Marriage" (PhD diss., University of Pennsylvania, 1918).

[253] *Propinquity* is a word that I'd never heard of until this past year, but in the course of this year I have heard it several times. Propinquity means nearness or closeness. It can refer to nearness in time, in person or place, or in relation, like kin. I first encountered this word when I stayed at a vacation beach house in Florida called Propinquity. I then saw the word on an elevator in a hotel shortly thereafter. I lastly came across the idea in reference to whom we marry in my research. We tend to

is that these principles sneak into our lives sometimes without us being aware of them.[254]

If you were to spend most of your life in a small town in a rural area like Appalachia, you would likely marry someone from a similar background and with the same patterns and cultural beliefs as you. That is great if that is your frame of reference and you are happy with the outcomes there. In fact, you may be likely to stay in the same geography for similar reasons, even if things seem challenging job-wise, relationship-wise, money-wise, or other.[255] We are comfortable in the environments that we have always known, and it is easy to simply stay and accept these conditions, rather than make more effort to go somewhere that is unfamiliar. The same behaviors and choices that we make in terms of our geography therefore can similarly affect our choice of a spouse. As we know this may be for cultural reasons (e.g., families of certain cultures cluster geographically together) or due to race clustering, religious affiliations, or other national ethnic congregations (for example, Filipino people have a tight-knit community in a nearby town to my Appalachian community and encourage marriage within their own culture).

This idea of propinquity is important if you want your kids to choose mates of similar balanced backgrounds who enjoy similar cultures or who have similar levels of education or similar ambitions. As I said earlier, I picked a less efficient area in geography in terms of our state for other reasons, but marital choices and related outcomes illustrate this as well. I am not originally from this geographic area, so it is naturally more difficult

become attached or attracted to people we are close to in geography, like a few blocks from where we live, at the workplace, etc. Propinquity is a powerful driving force sociologically.

254 I would also point out that propinquity needs to be carefully considered after marriage, as you may become attracted to people in frequent proximity to your workplace or otherwise, which is sometimes a factor for affairs during marriage. Hence, this could lead to divorce in much the same way if you are not aware of this invisible principle.

255 I keep asking myself why people continue to stay in the poorest state in the country with the worst well-being index, worst health outcomes, and the worst educational levels, and I keep coming up with this idea of propinquity. It is almost like an internal homing device that draws you back to and keeps you in your place of birth or origin. It is ironic that I am now finding myself doing the same thing; I am moving to a new geography less than a few hours from my own birthplace, and while it is more than fifty miles, it is surely close. Coincidence?

for me to accept unusually intolerable imbalances in an area in which I feel no natural connection, as the idea of propinquity would otherwise govern.

West Virginia, the state where I've lived for the last fourteen years, has had more than its share of bad outcomes that are symptoms of a bigger problem, which concerns me as a father. West Virginia has an unusual prevalence of perpetual poverty, coincident with the most uneducated population in the country who favor a large amount of welfare (greater than 20 percent), a substantial lack of employment,[256] and one of the highest divorce rates in the United States (third highest[257]), as well as worst overall sense of subjective well-being in the nation.[258] This is simply not the ideal population for my girls to choose a spouse from. It is actually a potentially toxic mixture.[259]

This is one reason I've facetiously suggested they consider the MRS degree (especially in college or graduate school but preferably in another state since even West Virginia's colleges do poorly in many regards[260]), but after much research and thought, it does not seem so unrealistic to look for your spouse in college or graduate school, especially considering the frame of reference.

For my children, I sincerely think that type of person will more likely meet their expectations and match their lifestyle desires (in terms of understanding the need for some serious effort rather than a little or of matching a minimal level of effort with a productive outcome) more so than someone in their local geography who does not have similar life goals

---

[256] Only six of ten people are employed in West Virginia.

[257] CDC/NCHS, National Vital Statistics System from 1990–2011.

[258] We ranked fiftieth of fifty states for well-being index in 2011–2013. Gallup-Healthways. The well-being index includes topics like physical and emotional health, healthy behaviors, work environment, social and community factors, financial security, and access to necessities such as food, shelter, and health care. http://info.healthways.com/wbi2013.

[259] Having said this, there are some people that are salt of the earth that I would be thrilled to call my future in-laws. A good local example is the D. R. family. The father is not a doctor but works with one (me), and I would be thrilled if any of my girls were lucky enough to marry one of his boys. But they are an exception (and often already taken) and not the rule, and that is my concern with our choice of geography. The pickin's are slim, and numbers matter a lot in the world of probability.

[260] In one survey, WVU ranked 170 out of public institutions of higher learning, and in another it was not ranked at all (referenced earlier). The ROI for most schools in West Virginia are miserable. http://www.payscale.com/college-roi/full-list/by-state/West%20Virginia.

of ambition or similar models of reference from which to learn.[261] If my girls wait, they will also be older and wiser because they have chosen to go to college and because of their physiological brain maturity, which is also temporally skewed.[262] I realize this may sound rather elitist, but it is simply a matter of the inherent bias of objective numbers and physiological maturity.

God knows I love our small town of 3,800 people, and it truly is a "cool"[263] town, but to use a rural Appalachian phrase, in it "the pickins are slim" if you are looking for a mate (and even more so for one who is highly educated and has similar religious values). Let me show you the numbers if you were a high school graduate (the norm) and were considering marriage within the local geography.

In a small rural town of 3,800 people in West Virginia,[264] only about 6 percent of the population are high school age (228 people) at any given time, and of those, less than 20 percent will go on to earn a bachelor's degree (about 45 people). Of those, 20 percent will achieve some higher education or high achievement (by my criteria of quality values including religion, ethics, morals, education, family, etc.) and lead the kind of lifestyle to which my children have become accustomed (9 remaining). Of these, roughly half are boys (4) and half are girls (5), so the odds as you can see are vanishingly small that my girls will meet someone that will appeal to them in our geographic locale (only 4 people roughly out of 3,800 from a given year) who would be a reasonable match in their age range. This is

[261] Having said this, two of our dearest friends grew up together here and met in high school and have been the model couple for blissful marriage (much like Juran, whom I describe more in part 3) for twenty-five years, as were their parents. It can happen, no doubt about it, but the numbers make it difficult due to our changing values.

[262] Developmental maturity through myelination of the brain follows a directional time-related maturity that is skewed with the frontal lobes last to develop. The frontal lobe areas that develop at twenty to twenty-five years of age involve impulse behavior, emotional maturity, and our ability to plan ahead and think maturely. Until then, the brain is literally a moldable work in progress, as described by many experts in neuroscience. Therefore it only stands to reason to wait until after this time period to consider something as serious as marriage. OECD, *Understanding the Brain: The Birth of a Learning Science* (OECD Publishing, 2007).

[263] This is in reference to our designation as "the coolest small town in America."

[264] This is the estimated population of the town we inhabited for the past fourteen years.

where the power of geography (particularly in terms of population[265]) can make a disproportionate difference.

I realize this sounds rather snooty, but I can only tell you my opinion and how I think in terms of numerical probabilities. We all want the best for our kids, especially in marriage. I can say without hesitation that I want *more for less* here. I want the most happiness for my children with the least number of spouses for each child (i.e., only one is the preferred outcome for each in this measure, but the reality is one or more of my children will experience divorce on a statistical basis).

In a larger metropolitan geography like New York City (where I coincidentally first met my future wife) or Boston (where my sister met her future husband), the numbers are about 2,000 to 2,500 times higher, so there are nearly *ten thousand* potential matches instead of only four! Not only does this improve your odds of meeting someone socially more like you than not, but you can also be more selective and less impulsive due to the competing factor of a ticking biological time clock.

I personally want to optimize the odds for my children, and in order for that to happen, a person needs to take the time and do the work to make sure there is a measure of compatibility with a potential mate.[266] I would rather my children enjoy a mate from a population with ten thousand options rather than among a choice of only a few, especially knowing the disproportionate relevance of the long-term outcomes.[267] And while I would

---

[265] This is one of the problems West Virginia has. I have adjusted for the population by giving rates as percentages. If and when you have a small population, it can be difficult to achieve certain levels of progress due to lack of large enough funding or manpower. There is likely a critical mass that becomes necessary, and then growth becomes exponential or nonlinear as a function of itself. West Virginia has zero cities with population over one hundred thousand, and West Virginia is in the bottom eleven states (almost the bottom 20 percent by population rank). I honestly think this contributes significantly to West Virginians' outcomes.

[266] Likewise it requires effort after marriage to keep it working. The concept of least effort has to be continually realized and sometimes resisted, as it is a natural tendency (even in marriage) once we become content. It is always easier to quit a relationship, but making the effort is usually worthwhile since you probably will just repeat this behavior again in another relationship.

[267] I used conservative numbers here to give you the idea. Actually I looked up the educational attainment by state, and West Virginia was the lowest in the country for bachelor's degree or higher (17.3 percent, which was dead last; the average for the entire United States was 27.9 percent). For higher advanced degrees only Arkansas had a lower rate with Arkansas at 6.3 percent and West

not advocate trying out each one of these ten thousand choices one by one, I think the reality is the outcome can be narrowed significantly and chosen efficiently[268] by this power-thinking methodology and by simple numerical probability.

In a similar way, it makes sense logically to improve your odds by waiting until college (where the average college size is roughly twenty thousand to thirty thousand students) to meet your life mate, where the odds favor that you will find someone of similar ambition and similar life goals. In this example, roughly half are boys and girls, so there are again about ten thousand to fifteen choices instead of merely four for the average state college. And sure, if you really wanted to be super-selective, you would choose someone from a school that has a good track record of success, as a recent graduate from Princeton suggested.[269] All these choices would naturally improve your efficiency in finding a mate more like you in compatibility[270] and hopefully less likely to leave you.

Virginia at 6.7 percent; the average for the United States was 10.3 percent. US Census Bureau, Pub. 2012, Using 2009 Data, Table 233: Educational Attainment by State 1990–2009. In case you are interested in the trends, West Virginia has been last since 1990 (twenty-three years). Why you ask? Good question.

268 A few websites have in fact jumped on this idea and have been successful by increasing the number of potential matches with efficiency and compatibility measures (e.g., eHarmony.com and ChristianMingle.com).

269 Susan A. Patton (class of 1977 from Princeton) wrote about this in an editorial for the *Daily Princetonian* and had a good point in terms of matching intellect in her case with a mate of equal or better value. I think you need to consider your spouse as an equal in the modern societal relationship, or your relationship is doomed from the start. I would point out that Princeton has one of the highest graduation rates in the country (96 percent) as well as one of the highest starting salaries on average and top twenty net return on investment (ROI) for the price. http://www.payscale.com/college-roi/full-list/by-state/New%20Jersey.

270 This is another ability that Pareto would suggest follows these rules of imbalance, so pay attention to your *ability* here as well.

"Grandmother ... When Shall I Marry? This Year, Next Year, Sometime, Never."[271]

The Imbalance of Time in Marriage and Divorce

Some would scoff at the idea of getting the MRS degree in the modern culture, but during the early part of my childhood and probably even disproportionately more so during my parents' generation, the idea of getting the MRS degree was indeed very popular, as were other old-timey beliefs that we no longer value (like taking care of grandmother at home rather than in a nursing facility).

During the 1950s, women felt a lot of societal pressure to focus their goals on obtaining a wedding ring, since women were not as prevalent in the workforce. Today with equal opportunities and growing gender equalities and, more relevantly, the skewed effect marriage can have on your life, getting the MR degree is just as important. In other words, marriage should be thought of much like our education: it should involve preparation, time planning, lots of work, and other quality measures if you want it to succeed.

The US marriage rate in my grandmother's time (post–World War II) was at an all-time high, and couples were marrying on average younger than ever before (the median age for marriage was twenty for girls). Getting married directly after high school or while in college was considered the norm. A common stereotype then was that women went to college only to get an MRS degree rather than a bachelor's degree.

While I am not implying all marriages during this time period resulted in happier or more-successful outcomes, the general trend of happiness and better well-being in America did correlate with higher rates of marriages,

[271] This is from another version of "Tinker Tailor." There are many versions other than the one I used.

due in part to what I think were different values regarding marriage and our cultural efforts to remain together despite life's many hardships.[272]

Although women had other aspirations in life (e.g., careers), the dominant theme promoted in the culture at the time was that a husband was much more valuable to a young woman than a college degree. If a woman wasn't engaged or married by her early twenties, she was in danger of becoming an old maid. This is likely one of the reasons for the earlier marriage distributions in the curve I show in a few paragraphs. The curve has shifted to later years as the generations have changed in their values and in the modern opportunities available for women.

Nowadays, the idea of the MRS degree may be less in vogue, as women have become more independent and career-oriented in their own right, and marriage has taken a backseat on the time curve, but I can assure you marriage clusters in geographic and temporal ways still.[273] In fact marriage is delayed more now in both sexes as a result of young adults focusing on education and less on family creation until education is complete. While I don't truly believe that fewer people entertain marriage, I do believe the younger generation sees the older model as inefficient and nonproductive, particularly given the high rates of poor outcomes with divorce rates being so high in the modern era. I think it also has something do with individuals obtaining some degree of financial freedom before committing to a lifelong relationship with someone else, particularly having seen the model of increasing failed marriages from their parents.[274]

---

272 Now it seems that we are more about pleasure and selfish desires. We are more impulsive now and less inclined to do the work necessary to make a marriage work and last because society has simply learned that higher divorce rates are okay. Marriage means something different now than what it represented then.

273 A coworker of mine told me her mother's goal (in West Virginia) was to marry a man with a job (no joking). That was the extent of it, which gives you an idea of the frame of reference in a poor state with high levels of unemployment and disability.

274 The ideas of marriage and money as codependent variables comes from the observation that marriages plummeted after the Depression in the 1930s, surged again when times were good, then fell again during World War II, then surged again when the economy recovered in the late 1940s and 1950s. Since then it has been on a steady decline. Interestingly, as I show with newer data, marriage actually increases household income, independent of whether the added spouse is employed or not. In other words, being married somehow enhances household income separately from the addition of the spouse's income.

These are important concepts. I personally believe that couples that enjoy two incomes have less marital conflict, as money matters are one of the most common causes of strife in a relationship, and having some income-earning autonomy helps to generally raise the quality of the couple's overall lifestyle. I would likewise encourage all my girls to plan lifelong careers independent of marriage, as I think they will be happier in the long term using their own tool sets to obtain the lifestyle choices to which they have grown accustomed, independent of another person's influence. In the perfect situation where the marriage is also a good choice and lifelong match, these then become synergistic outcomes,[275] rather than simply additive from the sum of their individual contributions.

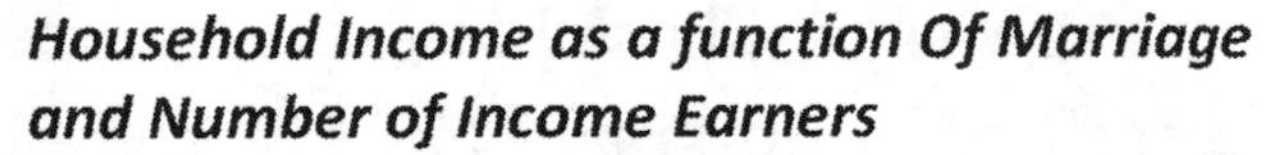

[275] This may be the nature of the math of these imbalanced outcomes—people that make multiple good choices have supra-additive outcomes, also known as synergistic, rather than simply the sum of each. For example, in the case of a synergistic outcome, 1 + 1 = 4 instead of 2. "Societies are rarely ever the simple sum of their individual components." Pareto, *The Mind and Society*, 1:32.

This image shows the effect of marriage on household income, which makes sense to me. Being married is almost twice the efficiency of being single in terms of household income, but a lot of this comes from the spouse earning more income as a member of the paid workforce (especially in today's America). Still, there is a separate synergistic effect that results just from being married, and why that is would be pure speculation on my part.[276]

When people choose to get married follows a typical time-dependent pattern that is also skewed, based pragmatically on simple biological principles (our reproductive years limit the biological clock to a span of about twenty to twenty-five years) as well as other factors to which I have already alluded. What is interesting to me, however, are the recent generational shifts in later marriages, based I would assume on the fact that more women are entering the workplace and are therefore delaying families until they are ready from an education and career standpoint. I think it also relates to the relaxation of societal values (e.g., the sexual revolution) since the 1940s. But I also hypothesize that this shift suggests a learned efficiency in choosing the right life partner, which naturally takes more time (and yes, more effort) and greater exposure to potential partners. I bring this up for relevance here since I have three daughters that these ideas will potentially affect. [277]

---

[276] US Census Bureau, Table HINC-05: Percent Distribution of Households, by Selected Characteristics within Income Quintile and Top 5 Percent in 2011. While I did not show these curves separately from the number of contributing members, you will have to take my word on this idea that marriage as a separate variable adds to income separately.

[277] My Alma Mater (UVA) has a research initiative called the National Marriage Project with great data and resources on marriage, which is where I obtained some of this information. University of Virginia, *The National Marriage Project*, http://nationalmarriageproject.org/. The original data came from Rutgers by way of the principal investigator, David Popenoe, who, at the time of this publication, is at UVA. The dotted horizontal line is roughly the median household income for Americans.

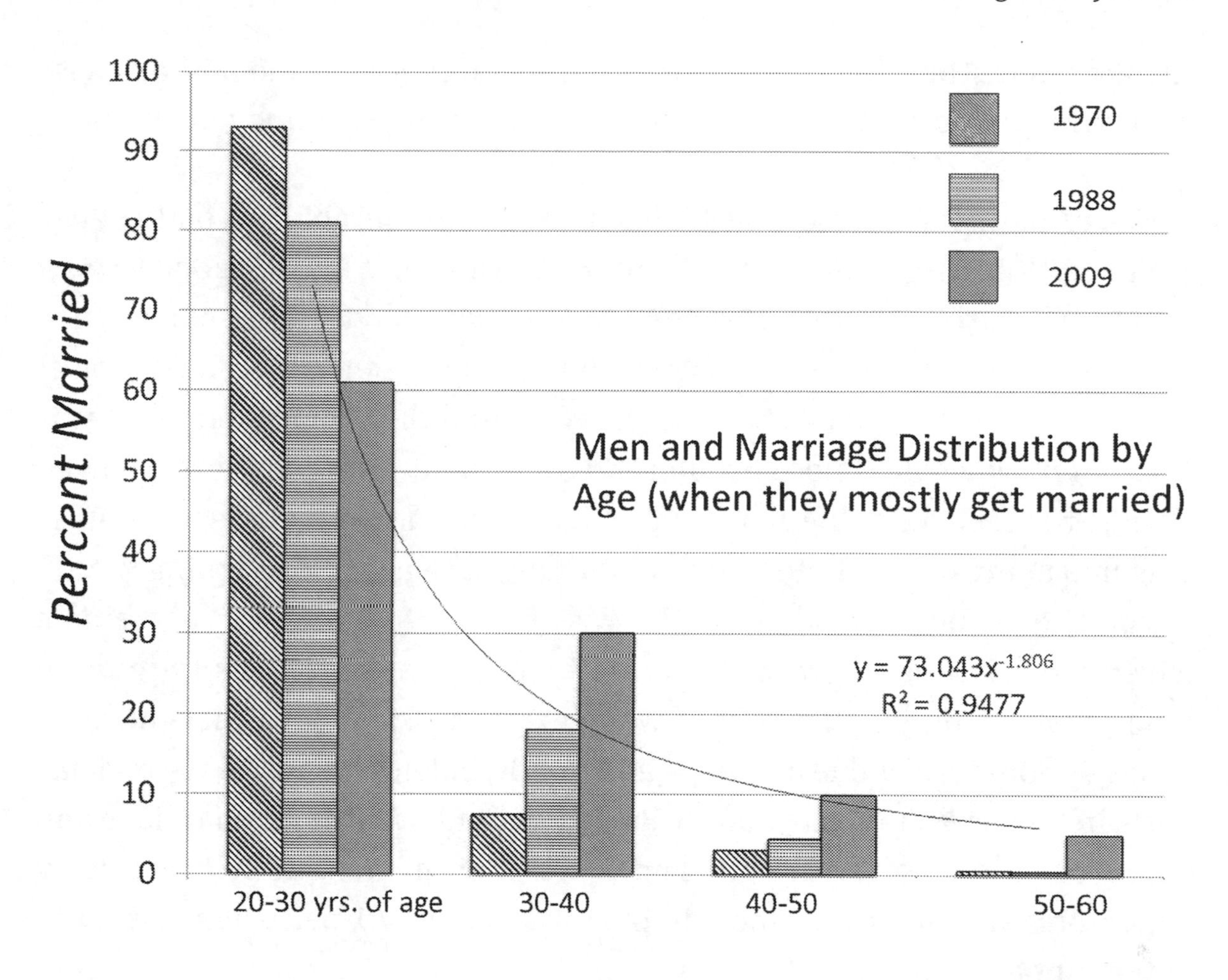

Marriages (for both American men and women[278]) are skewed in their distribution, much like the imbalanced distribution for income (it appears to be a power-law rank distribution). Over time, the marriage curves have shifted over toward the right to later ages (for men the median age for getting married for the first time is 28.4 years now versus 22.5 years of age in 1970). I personally think this is a good thing. I did not show the curves for women here, but they are similar (for women the median is now 26.9 versus 20.8 years of age in 1970). The biological clock may be in part responsible for the efficiency of the temporal clustering seen[279] (in women this explains why the marriage age is younger than men), but it is nevertheless being

[278] I plotted the skewed distribution for men out of personal curiosity, although the curves are nearly identical for women (not shown). Diana B. Elliott and Tavia Simmons, "Marital Events of Americans: 2009," *American Community Survey Reports*, US Census Bureau, August 2011, 13–14.

[279] In this case I found 27 percent of the time periods accounted for 77 percent of the marriages, which is a Pareto distribution similar to income. This makes sense in regard to our biology.

pushed out farther to the tail due to other social concerns and perhaps with the help of modern technological improvements in pregnancy planning.

Delaying marriage beyond high school and preferably beyond college is a good thing in our present culture, and although it may be good to wait to find a future mate in college or graduate school, delaying marriage is good for many other reasons.[280] According to the National Marriage Project *Knot Yet* report, "Delayed marriage has elevated the socioeconomic status of women, especially more privileged women and their partners, allowed women to reach other life goals, and reduced the odds of divorce for couples now marrying in the United States." Waiting is clearly a good thing.[281]

In the time period that I was born, 90 percent of female "twentysomethings"[282] were married by age twenty to twenty-nine. In today's world, only 50 percent of women are married by twenty-nine.[283] I believe people still value marriage and family, but now they choose to delay both. In fact, most (approximately 80 percent) adults still say marriage and family are the *most*-important elements of their life (about 20 percent say it is important but not *the* most important, and only 1 percent say it is not important at all).[284]

---

[280] Malthus argued that two types of checks hold population within resource limits: *positive* checks, which raise the death rate, and *preventive* ones, which lower the birth rate. The positive checks include hunger, disease, and war; the preventive checks include abortion, birth control, prostitution, postponement of marriage, and celibacy. *Wikipedia*, s.v., "Thomas Robert Malthus," last modified August 5, 2014, http://en.wikipedia.org/wiki/Thomas_Robert_Malthus. In later editions of his essay, Malthus clarified his view that if society relied on human misery to limit population growth, then sources of misery (e.g., hunger, disease, and war) would inevitably afflict society, as would volatile economic cycles. On the other hand, preventive checks to population could ensure a higher standard of living for all, while also increasing economic stability. I am not a proponent of his theory in total, but he has some valid points.

[281] Kay Hymowitz et al., *Knot Yet: The Benefits and Costs of Delayed Marriage in America* (The Relate Institute, the National Marriage Project at the University of Virginia, and the National Campaign to Prevent Teen and Unplanned Pregnancy, 2013), 3. My kids have of course heard the lecture about waiting for sex, but I realize that fell on deaf ears for the most part. The same goes with delaying marriage, and I hope to be more efficient (100 percent) in this message. I likewise want to be more efficient with the idea of delayed childbirth to a degree, though waiting too long poses other negative outcomes that must be balanced as well.

[282] This is a term used by the authors from Knot Yet.

[283] Ibid., 12.

[284] Ibid., 14.

Women economically also enjoy an annual income premium if they wait until thirty or later to marry. For college-educated women in their mid-thirties, for example, this premium amounts to $18,152 per annum.[285] Over a lifetime of potential working years, this translates into almost $1 million of extra income, which is the same difference seen with different levels of educational attainment (even above college) and with efficient matching of majors within college! In theory, they are all additive inputs. So if you are so inclined and need or want to work, which seems to be the trend with modern American women, take note of these facts.

I can already anticipate problems and complaints from readers.

How can I encourage my girls to think about marriage like an occupational choice and simply ignore the attractions between people that produce a natural "chemistry" that should not (or cannot) be analyzed? Well, my scientific answer is that "chemistry" through molecular attractions[286] obeys the same power-law principles (of attraction and repulsion) and can be studied objectively. Chemistry between molecules and chemistry between individuals are just different scales of similar phenomena—they may be responsible for the physiological reason for the social law of attraction, so your relationship should take this into consideration as well. Why should marriage be approached any differently than everything else in the universe? It is simply another choice with skewed consequences like everything else.

I want my girls to be prepared by knowing these natural laws and relationships, and I really want them to spend the time, energy, and effort to make sure that they are compatible with the person they marry so that they don't become one of the bad statistics I discuss. I am not implying my children cannot be attracted to someone for other reasons—there is a certain chemistry that needs to be explored and experienced—but when

[285] Ibid., 15.

[286] Van Der Waals forces of attraction are the molecular equivalent of the laws of attraction, as well as electromagnetic repulsive and attractive forces (or for that matter gravitational attraction between two masses). All these forces follow the same principles, creating a natural assortment driven by power functions and power-law distributions.

it comes time to get serious and think about committing to a family, they must learn to really examine these issues critically.

Those who don't choose wisely experience the disproportionately negative consequences of their bad choices and all the corresponding spillover into personal well-being. With marriage, the negative side is of course divorce. As most of us know, divorce is more culturally accepted today than it was in previous generations. It almost seems like the cultural norm. If things are not working out and two people are not compatible, then by all means quit the relationship, right? For any problem, quitting is the easiest solution and involves the least effort after all! The current estimates in fact support the fact that divorce rates are nearly 50 percent as a rule.

Americans perceive their marital relationships now in the same way we think about the upkeep for our cars: we trade the older models out for newer ones when the effort to keep them up becomes more involved than we are willing to commit to long term. This gets back once more to our changing values and our willingness (or lack thereof) to honor commitments (often requiring more than some minimal effort) we have made before others and especially before God.[287]

One quick way to see your life change radically and often for the worse (and go from relatively rich to poor) is to see your family become imbalanced in marriage, which affects financial security, retirement, and overall lifestyle (i.e., happiness and success and well-being). And the easiest way to see this all evaporate is to get a divorce. A friend of mine, who is a very good (and busy) divorce lawyer in our small West Virginia town, will tell you that this is a lucrative business for the lawyer in today's world and is a quick way to reverse your good fortune and success while improving his. Divorce usually results in a rapid decay of your assets and your sense of well-being.

[287] I am no more perfect in this regard than any other. I have been close to divorce with my wife on two occasions in twenty-five years, and what helped me more than anything was a lot of effort (including marital counseling), friendly and timely reminders (and prayers) from my pastor and friends about the commitment I made to God, self-reflection, and my willingness to forgive in the relationship. It takes two to tango though, and we cannot control our spouse in making the same efforts.

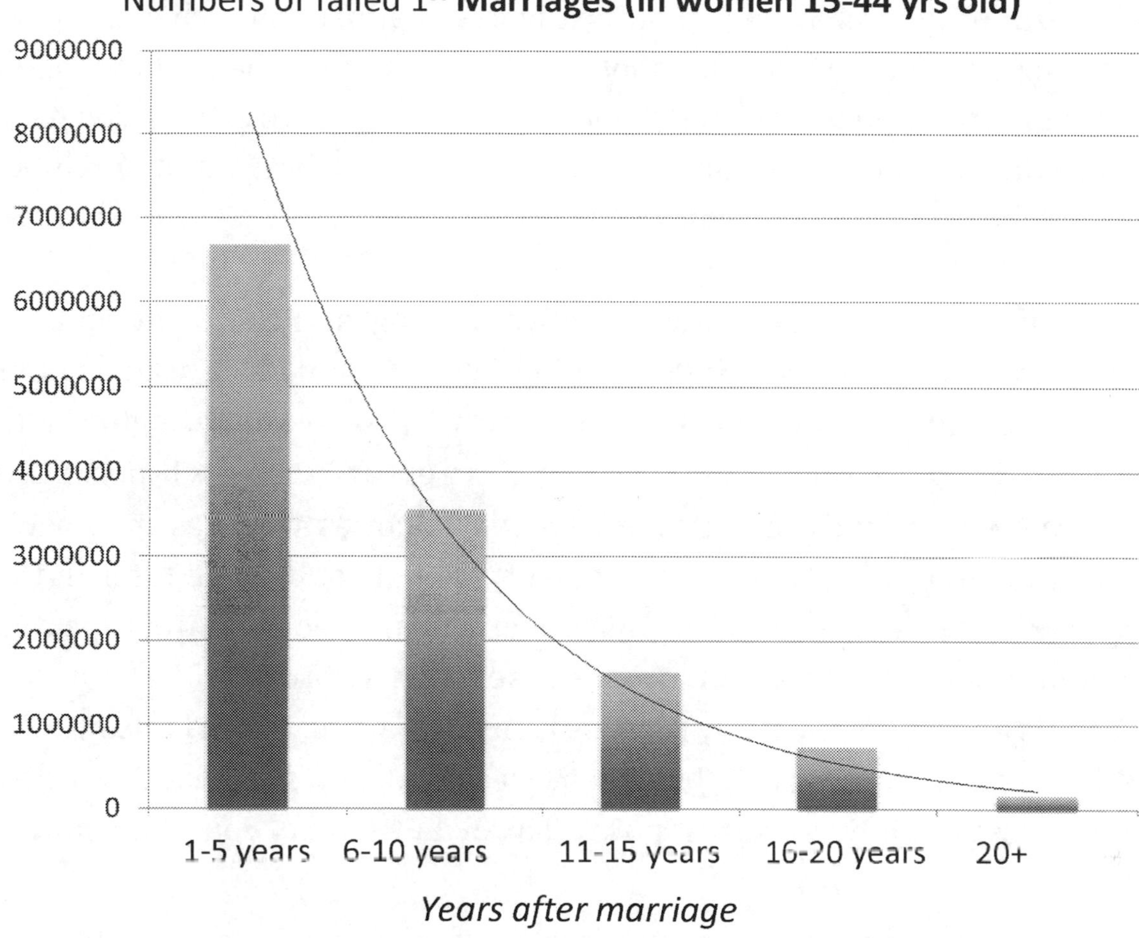

Marriage failures (i.e., divorces) follow a similar skewed or asymmetrical distribution with time. Divorce rates peak temporally at three to five years (most people quit early) following marriage and then decay from then on.[288]

The temporal distribution of divorce is a function of many variables, including the age when you get married, your income, your educational attainment, ethnic background, and so many factors, many of which are similar to what's seen with income distributions and households. Translated, this says a lot of people get divorced early, and fewer get divorced as time

---

[288] Matthew D. Bramlett and William D. Mosher, "First Marriage Dissolution, Divorce and Remarriage: United States," Advance Data, number 323, Centers for Disease Control and Prevention, National Center for Health Statistics, May 2001, 1–19.

goes on.[289] Stated another way, most of the divorces are accounted for by 20 percent of the time, and few of the divorces happen in much later years once you pass some critical threshold. It looks to me as if the first five years are the critical time period, and if you can get past this threshold, you have better odds for success in your marriage if you are willing to put forth the efforts required to make it work.[290]

Marriage and divorce rates also vary geographically and cluster socioeconomically, as you might imagine. West Virginia as a state is poor and uneducated overall, and these inputs tend to correlate highly with many other negative outcomes, such as high stress and divorce. I speculated marriage would follow money but apparently not as much as divorce.[291] As I anticipated, West Virginia has one of the highest divorce rates in the country.[292] Once more, West Virginia is trapped in a geography of repeated unhappiness with the measure in this case being divorce.[293]

This is another of the many reasons I am now raising my girls elsewhere. Yes, I know that sounds petty, but when you combine these ideas all together and understand how powerful power laws can become, you will come to

---

[289] US Census Bureau. USCensus.gov. I might also point out that a disproportionate number of divorces (80 percent) come from repeat offenders (20 percent of the divorcees).

[290] Recent data also shows that for people older than fifty divorces have increased from one in ten in the past twenty years to one in four now. Susan Brown and I-Fen Lin, "The Gray Divorce Revolution: Rising Divorce among Middle Aged and Older Adults 1990–2010" (Working Paper Series, WP-13-03, National Center for Family and Marriage Research, Bowling Green University, March 2013), 2.

[291] West Virginia also has a disproportionate number of lawyers I think for this reason. A good friend of mine is a lawyer, and he ranks divorce law as one of the top reasons for client referrals in our area. He would be an efficient choice for Daisy because he is the lawyer, and he sees the consequences of people's stupidity on a regular basis. Unlike me, he loves stupid behavior in humans, as it puts food on his plate.

[292] West Virginia is third highest in state divorce rates at 5.2 per 1,000 population (tied with Oklahoma), which ties in with poverty I am certain (the number one cause of marital conflict is finances). The only two states worse were Nevada (at 5.6) and Arkansas (at 5.3). Data from Center for Disease Control and Prevention, 2011. West Virginia's average over twenty years was 5.1 per 1,000, which is consistently among the worst. I don't mean to imply the state causes the divorce. Rather, I think the high divorce rate is related to the poor repeated conditions, like poverty, low education, and other poor variables that cluster in our geography so efficiently. Incidentally America is also higher than most other countries in this regard also. After the writing of this book, my family and I have moved to North Carolina, which has a rate of 3.7 per 1,000 population (29 percent relatively lower).

[293] Divorce Rate by States 1990, 1995, 1999_2011, http://www.cdc.gov/nchs/mardiv.htm#state_tables.

the same conclusion that I have (unless propinquity trumps intelligence for you).

Unfortunately poverty and uneducation begets more family strife and divorce, which then propagates more poverty and ignorance, and it becomes a vicious downward spiral toward the eventual outcome of *more stress, more divorce,* and *more unhappiness*—all bad outcomes prevalent in West Virginia and other geographic clusters.

For the United States as a whole, *only 63 percent* of American children now grow up with both biological parents. This is the *lowest* figure in the entire Western world! This is very imbalanced and is a problem in our family value system domestically. In the near future, nearly 50 percent of children will at some time grow up in a home with a single parent.[294]

I think these issues are major contributing factors in the growing measures of our general unhappiness and poor well-being as a nation. I explore this idea of how the family unit has fundamentally changed in the final parts of this book.

In addition to taking the appropriate efforts in choosing your spouse (and the timing of marriage), you should likewise think about optimizing family size (in terms of costs[295]) and try to plan your family at least to some degree. The number of children you have and when you have them can affect your lifestyle considerably (likewise the number of people living with you and consuming resources and not contributing resources is a factor) and is more simply a matter of responsible behavior. Recall the last line of the nursery rhyme that has us consider how many kids you will have: "one, two, three, four ..." This is an important variable in your lifestyle consideration. *Four* is probably too many children to support in today's modern world.

---

[294] National Marriage Project and Institute for American Values, *The State of Our Unions* (Charlottesville, VA: National Marriage Project and Institute for American Values, 2005); Elizabeth Marquardt, David Blankenhorn, Robert I. Lerman, Linda Malone-Colón, and W. Bradford Wilcox, "The President's Marriage Agenda for the Forgotten Sixty Percent," *The State of Our Unions* (Charlottesville, VA: National Marriage Project and Institute for American Values, 2012).

[295] Wealth is assets minus debts or costs. If you have average to low income, you will never achieve any comfortable wealth if you reproduce too often. The costs disproportionately rise with no change in the other inputs. This is another learned inefficiency within our impoverished population, in part propagated by the welfare system and the increasing rewards it offers to have more children within already poor families.

One to two is probably just about right as an average, but if you have the means and the desire as my wife and I did, then more power to you! But please don't make this decision if you don't have or anticipate the means to provide the necessary support.

Experts estimate the cost of raising a child in the United States is roughly $250,000 over a lifetime.[296] Two children would cost perhaps a bit less than $500,000 due to some overlapping costs, but in general the costs rise dramatically as you have more children. This is true whether you have them biologically or adopt them. Our rationale for having four was that we had the means to support them (based on good occupational choices), wanted more, and felt it was our obligation to provide for more since we had the means. We felt adoption was a way to provide for what was already out there in the world (once more I was thinking efficiently). Our two adopted girls are now indistinguishable from our two biological kids with the exception of their genetic composition. But the point is to think about all this in the planning process and use good judgment or else be prepared to compensate in other ways due to the consequences of your choices.

And no unplanned pregnancies, girls, please!

The University of Virginia research group the National Marriage Project recently reported that one important consequence of delayed marriage in our country is that most Americans without college degrees now have their first child before they marry. By contrast, the vast majority of college-educated men and women still put childbearing after marriage. *Knot Yet*[297] explores the causes and consequences of this revolution in family life, especially the ways that delayed marriage is connected to the welfare of "twentysomethings", their children, and the nation as a whole.

---

[296] Mark Lino, "Expenditures on Children by Families, 2011," Miscellaneous Publication 1528-2011 (US Department of Agriculture, June 2012), www.cnpp.usda.gov/Publications/CRC/crc2011.pdf.

[297] The *Knot Yet* report is a joint effort from the National Campaign to Prevent Teen and Unplanned Pregnancy, the National Marriage Project at the University of Virginia, and the Relate Institute. twentysomethingmarriage.org.

In America, this concept of children born out of wedlock is one of several growing problems in the family value system that is soon to become the so-called modern family. In women under thirty years of age, 53 percent of childbirths were out of wedlock. Unwed childbirth is the new norm according to some.[298] There is a correlation with this outcome with education and socioeconomic class, as mentioned previously. More families with children born out of wedlock coincide with the disappearance of the middle class, along with several other values (e.g., education and achievement) important to Americans.

This new norm for America is predicted to result in a not-so-great outcome for our future generations. A new phenomenon, the Great Crossover, previously seen in the lower socioeconomic groups, is also now occurring in the middle classes. This refers to the idea that the median age of first childbirth (25.7 years) for all women is now less than the median age of marriage (26.5 years).

The top quintile (top 20–30 percent) of women in America (i.e., the upper class) based on income, however, for some rhyme or some reason, seem more inclined socioeconomically to delay and plan their childbirths. They may better understand the consequences of these choices, have been taught different values, or have learned something else. I could hypothesize these outcomes occur for many reasons, one of which is the need to delay children until after college or some level of career achievement, which seems logical and efficient, but the reasons are not as relevant to me as the clear understanding of the *outcome-choice* imbalanced relationship. This helps explain why we see a skewed distribution of marriages in America and helps explain some of the link we see with income imbalance within households as well.

Indeed, this negative crossover has not been seen in *college-educated women,* who delay their first childbirth typically more than two years after they marry. This is another reason I emphasize college for my Daisys—not so much to shop for a husband but to learn the value of delayed gratification, especially as it pertains to sex and family planning, among other things.

---

[298] Marquardt et al., "The President's Marriage Agenda for the Forgotten Sixty Percent," 2.

In fact, delayed marriage confers financial advantages as well as other favorable outcomes (general well-being for example).

If you had to guess which state enjoys the highest unwed birth rate in America (for Caucasians), what state do you think it would be? Yes, West Virginia won that title with a 40.8 percent rate of births occurring out of wedlock for whites (the highest in the country).[299] West Virginia doesn't really have much of a black or Hispanic population to speak of, so those particular stats are not pertinent here for my kids, and sociologists cannot therefore blame these figures on racial differences as is commonly used as an excuse elsewhere.[300]

This is another example where I don't think the environment I originally chose to raise my family is the most favorable for my girls. I trust them implicitly, and while I *assume* an unwed teenage pregnancy would not ever be a major concern, environment surely can play a disproportionately big role (i.e., peer pressure), as I have seen with my son. And I have furthermore learned my lesson about what I *assume.*

I currently have three girls, all lovely and all interested in marriage one day. Unfortunately they live in a country where in their future most children are unlikely to grow up with both parents and in a state that has the third-highest divorce rate in the country, as well as other potentially unfavorable outcomes.[301] Those are not optimal odds for a healthy environment of

---

[299] US Census Bureau, "Births," *Statistical Abstract of the United States: 2012*. The most recent stats I could find were from 2008 and for non-Hispanic whites, since we have negligible amounts of Hispanic Americans and African Americans in West Virginia; the same date is in table form at http://www.heritage.org/multimedia/infographic/2012/09/unwed-birth-rates-by-state. Likewise we have the lowest percentage of Asian Americans within America, which is another reason my adopted Asian American daughter wants to move away from here since she feels like such a minority. Overall, if you count all groups, Louisiana and Mississippi won these honors.

[300] West Virginia is very imbalanced in other ways with 94 percent white, 4 percent black, and 1 percent Hispanic population—another skew in West Virginia's ethnic makeup. So our legislators cannot blame the skewed outcomes on ethnic groups as is often done elsewhere. We can blame it in part on poverty though. US Census Bureau, 2012 data.

[301] Don't think that moving will get rid of this risk. Estimates from colleges show a prevalence of one-third of college women with HPV currently, and it will only get worse with this generation's lifestyle and behavior. Remember the vaccine is not entirely preventative, so use other methods (like abstinence or protection).

marital success for my girls or otherwise. This may be another invisible reason why people choose to live (or not to live) in certain states or cities (or even relocate to other geographies).

Interestingly the young generation within the state is efficiently getting the message in part, as recent reports from West Virginia show the working-age population from ages fifteen to sixty-four is declining (they are moving away from West Virginia for many reasons, none of which are *trivial*), and this trend is expected to continue at least another twenty years![302] These youth within our state that are looking for better well-being realize a sinking ship when they see it, and they are making the more-efficient choices since no one else will, including their parents![303] Sadly, without changes being made from within, there are no other choices sometime except to move elsewhere, which tends to further exacerbate the problem.

So in summary, this is what I recommend for children in America:

1) Marry when you are ready and after you have thought it out well and planned for it (later is better). Twenty to thirty years old is good. Put more than a minimal effort in your choices. Do your research!
2) Plan your family with similar efforts, thinking about number of children, where you want to live, when you want to have kids, and so on, remembering the power of geography and thc other variables I've discussed (like propinquity).
3) Once you are married, stay married by putting more than the least effort into your relationship (i.e., avoid divorce, which is a big factor for many negative outcomes). This makes your choice of spouse even more critically dependent on quality, according to Juran's quality

[302] While the youth population is increasing for the United States as a whole by 6 percent, it is shrinking by 8.2 percent in West Virginia, which is the state with the highest gap between the old and the young. Having one of the worst employment-population ratios in the nation (we were forty-ninth; Mississippi wins the poorest distinction) spells trouble for our future economy over the next twenty years.

[303] I finally recognized we had a problem in our geography when my kids were begging me to send them to schools out of state while they were in middle and high school. We have since moved to North Carolina and are blessed to be able to have this type of freedom of mobility.

principles (think about the *vital few* and the *useful many* factors you consider important in a spouse before you take the vows, and don't compromise these values thinking you might be able to change your partner).

4) Pursue your own career outcomes as you would were you not married, and adjust as needed once you are married.
5) Teach your kids to become independent (by having them get an education and read this book) and to place value on the sanctity of marriage.
6) Plan your kids' lives with their objectives in mind, and integrate what you are able into your life plan (helping them with educational efficiency, college costs, etc.).

"Tinker, Tailor, Soldier, Sailor."

The Imbalances of Education within America

As I alluded previously, education is severely imbalanced within America and by no coincidence similar to the skewed income imbalance I discussed in part 1. A *tinker* requires more education than a *tailor* who requires more than a *soldier* or *sailor* (although rank in the military certainly improves with higher education and improved skills).[304] My oldest daughter wants to be a tailor,[305] and my son currently aspires to be a soldier. They will both quickly come to realize that they need a minimum level of some education as well as a consistent valuable skill set (like an ability to think critically) in each of their respective choices in order to excel.

Sadly, none of my kids has aspirations to be a tinker (i.e., engineer or scientist). In fact few people in America want to be the tinker![306] The tinker—or, as I prefer, the *thinker*—is one of the least popular careers and least popular majors in college, but there is no question our society disproportionately favors this kind of critical thinking more so than manual labor. While we always need both types, critical thinking is what is mostly

---

304 As I point out elsewhere, the best return on investment for colleges and higher learning is often through military colleges.

305 My daughter wants to be a theatrical makeup designer/artist and has now realized she will likely do better with a college degree. My son has not learned this yet, but he eventually will. My other two are still too young to really know anything, although one thinks she wants to be a veterinarian (a doctor), and the other wants to be a dentist (another doctor). Three of the four therefore have concluded that college, as a minimum, is a major factor in their outcome desires.

306 I showed in another chapter that only 3–5 percent of our population is the tinker (I included engineers and architects and scientists initially in my analysis, but after adding in the modern computer programmers and mathematicians it goes up to 5 percent of the population). In this case, 16 percent of people are the folks Daisy alluded to in her rhyme (doctors, lawyers, merchants, chiefs, and tinkers), and 84 percent are not.

needed in the growing world of high technology and even within the military,[307] as I am certain my son will see!

The imbalance within our education system can be seen to positively correlate more with income than any other variable, and its distribution relates in my opinion to our self-estimated principle of least effort that Zipf described in his principle of human behavior. It is mostly a *choice* whether we consider finishing high school or not (often based on these principles),[308] and similarly it is mostly a choice whether we consider tertiary education (how we do it is another matter).

I am able to see this relation more clearly since I have lived in the state with the lowest levels of educational efforts and attainment in America[309] as well as the second-lowest income per capita and household income in our country (outcomes which I know are correlated). Therefore I have witnessed firsthand how the value we place on education and in achievement pervades popular culture. We are efficient at achieving poorly in both income and education for a reason, and I think these are self-perpetuating formulas for social and economic failure, whatever outcome you choose to measure.

West Virginia has long been a geographic cluster of low-tech, mostly manual labor, rather than high-tech, education-dependent occupations.[310] Coal mining, for example, historically played a large role in the early years of West Virginia's economy. Needless to say, this industry is not one that is dependent on high levels of education, partly explaining the nature of our geographic-related educational disparity, at least historically. Other industries West Virginia favors include farming (West Virginia has two to

---

[307] Thinkers in the military usually end up being the leaders (the military elite) and are happy to give orders to the majority non-thinkers. The military is notorious for its value attachment to rank and hierarchy, and the military always plays a role in the governing elite that Pareto considered so valuable.

[308] West Virginia was consistently in the bottom 20 percent of states for high school graduation rates for the periods of 1990–2009 despite high school being mandated.

[309] Of West Virginia's population, 17 percent achieve bachelor's degrees or higher, which is roughly half the US rate.

[310] West Virginia has fewer educational degrees as I showed earlier. In addition, if you look at occupations that typically require these—professional, scientific, and managerial professions—we are also lower (7.5 percent of our population) than the national average (10.7 percent). US Census Bureau, American FactFinder, census.gov. The only other industry I saw us differ in was manufacturing.

three times more farming and agricultural jobs than the national average[311]), food preparation, and similar low-education, service-related occupations. West Virginia also supports one of the highest rates of minimum-wage workers in the nation in part due to this disproportionally heavy weighting of these mostly under-educated occupations.[312]

Parents typically want better lives for their children and usually suggest their kids educate themselves out of this downward spiral in order to achieve better outcomes. This seems logical, but the message is not being delivered efficiently enough. Part of the problem is West Virginia is not pro-business (especially industries that favor college degrees) in many ways,[313] and part is West Virginia is not pro-education.

In this chapter, I want to show four related concepts and how they all relate to our educational choices through implied minimal efforts, using my personal choice in geography and my family as a model.

The Choice to Pursue More Education

Education within the rest of America is in truth not really profoundly different than in West Virginia. Education is in fact unequal and imbalanced similarly everywhere with pockets and clusters of uneducation skewed in many other states, often linked to the same social and economic barriers. While Americans do have more children now matriculating to American colleges on average, the challenge is getting them to complete college and helping to make it affordable (West Virginia's state flagship school, for example, has a miserable four-year graduation rate and a disproportionately

[311] In West Virginia, the numbers are 5.2 percent versus the national average of 1.9 percent in all of America.

[312] The states with the highest proportions of hourly workers earning at or below the federal minimum wage included Mississippi, Texas, Alabama, and West Virginia (between 9 and 10 percent). The states with the lowest percentage of workers earning at or below the federal minimum wage included Washington, Oregon, Alaska, and California, all at or below 2 percent. It should be noted that some states have minimum-wage laws establishing standards that exceed the federal minimum wage. Bureau of Labor Statistics, US Department of Labor 2012 data.

[313] CNBC recently ranked West Virginia as forty-eighth of fifty for worst places for doing business in 2013 for many reasons, which I discuss in the next section.

poor six-year graduation rate[314]). This gets back to the ideas of opportunity that President Obama has proposed for years. This serious problem in our country needs to be addressed if we hope to make efficient changes that will benefit more people.

If I were to plot the levels of educational attainment for the whole of America currently, by looking at high school levels and above, the skewed curve should seem very familiar to you by now.

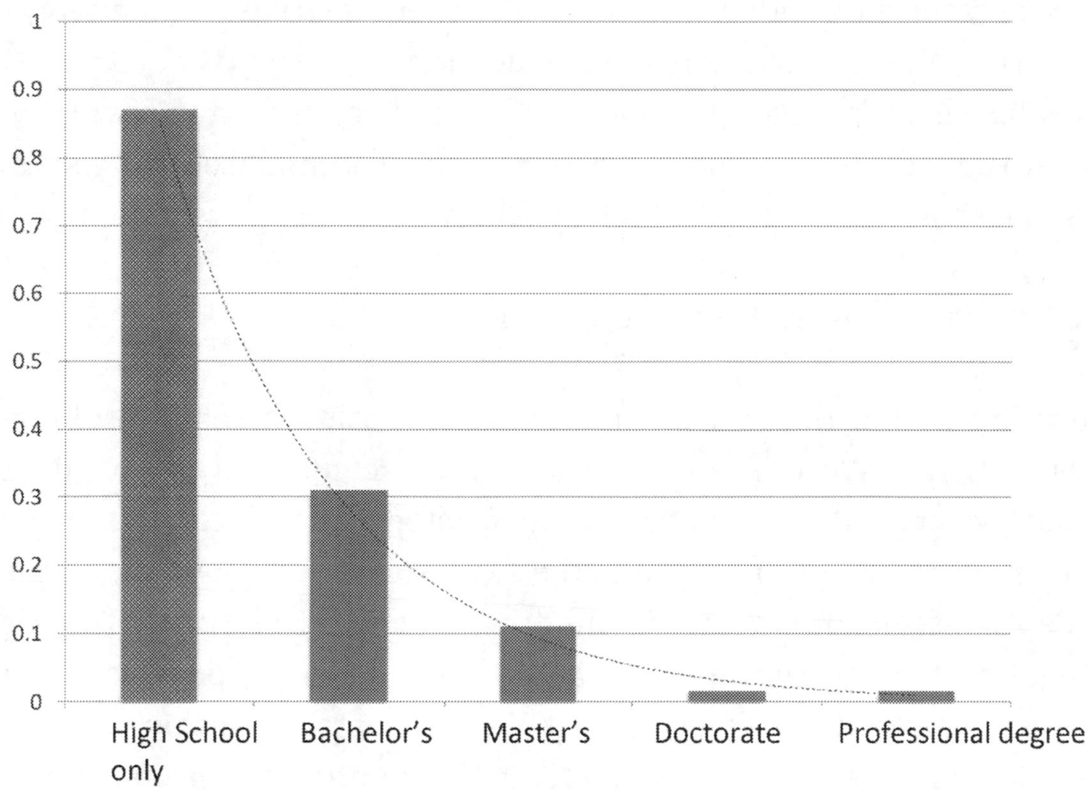

This graph shows the percentage of the American population with the various levels of educational attainment for 2012 (on the y-axis as an

[314] This shows how imbalanced these performance measures can be. UVA, for example, is consistently in the top tier with an 87 percent four-year graduation rate, which is excellent and is tied with Ivy Leaguers Princeton, Harvard, and UPenn (the highest in the report was 91 percent for Haverford, Swarthmore, and Pomona). WVU, the flagship university for West Virginia, has a 32 percent four-year graduation rate (and a six-year graduation rate of only 57 percent), which shows you a regional geographic contrast. "Colleges Report," *US News and World Report*.

average) versus the degree of achievement (on the x-axis). Less achievement is clearly more popular[315] and requires the least effort as an outcome, and these results show the nonlinearity (it is in fact exponentially decaying)[316] of America's educational outcomes.

Educational attainment is obviously one of the areas where we really suffer an imbalance as a nation. As you can see from above, the more popular outcome is the one with *the least effort*—high school graduation, for which we consistently achieve rates of 80–90 percent in the newer generation. This comes from our emphasis on the need for a high school education that began in the early part of the twentieth century and has now plateaued (we have maxed out on the efficiency there). In the 1950s, for example, we had only 50 percent graduation rates for high school. We obviously have learned to value high school as a minimum. The problem we have now is that a high school diploma is no longer sufficient for the demands of our industries. It is the modern-day equivalent of no degree. The new minimum for the twenty-first century is now the college degree.

My son unreasonably expects to make a six-figure income, but he has neither the skill set nor the education to do so. A high school education in general does not correlate efficiently with the top 20 percent of careers that I showed previously, so it is illogical to expect large income as an individual without appropriately higher education (or otherwise valued skills).[317]

---

[315] This data is from the US Census Bureau from 2011 and represents the average results for each category. Were you to choose to examine educational attainment by quintiles as I did income, I think you would see a clear correlation with income as well.

[316] US Census Bureau, "Educational Attainment in the United States: 2012," http://www.census.gov/hhes/socdemo/education/data/cps/2012/tables.html. I would be remiss if I did not mention there is a slight problem with the data in that this reflects the entire population and not what happened in one year but the results averaged into the population. This means it will be hard to show a significant change over time if the starting point is already low. What would be more interesting would be to track the changes each year to see the message and how efficient it is being delivered.

[317] US Census Bureau, "2010 Household Income," Current Population Survey, *2011 Annual Social and Economic Supplement*, http://www.census.gov/hhes/www/cpstables/032011/hhinc/hinc01_000.htm. Data is the number of households making the levels of incomes as a ratio of bachelor's or higher over high school graduate levels. As the desire to make higher levels of income increases (your threshold), then you will correspondingly increase your odds dramatically with a bachelor's degree or higher.

This is an example of one area in which he has mismatched expectations between what he desires and what he has achieved as an individual (and where I failed as a parent to teach this concept efficiently).

To demonstrate the influence of education on income success, if you want $100,000 a year of annual household income (this amount roughly represents the top 20 percent, then having a college degree disproportionately increases your odds of getting there. If you are interested in making $100,000–$125,000 per year, then you are two and a half times more likely to get there with a bachelor's degree or higher. If you want to make $200,000 or more, you are ten times more likely to get there with a bachelor's degree or higher. Ten to one odds are quite significant if you ask me.[318] Stated another way, for every ten people that get there with a college degree, one may get there with just a high school degree. The odds keep getting higher in favor of more education the more income you desire.[319]

I consider a high school education alone truly insufficient in an economy that is driven by heavy industrialization and technology, unless you are a genius[320] with great novel ideas like Bill Gates or Steve Jobs—there are exceptions to every rule[321]—or you are content with your manual skills as

---

318 I am not a betting man, nor do I play the lottery. But if you told me I had ten-to-one odds of winning at anything, I would keep doing it, and I would be happy with the outcomes almost always, especially the more I persisted.

319 I saw this in another way when I looked at income in West Virginia, which is among the lowest in the nation. Since we have *the* lowest educational attainment in America, a lot of our income issues stem from this factor alone. In fact West Virginia has more percentages of all incomes up to $50,000 (in other words, more people making the low ends of income), and then it declines quickly from there. It becomes more rare to make money after about $50,000–$75,000 as a rule for a household in West Virginia. The percentage of households that make $100,000–$150,000 is 8.5 percent in West Virginia (versus the norm in America of 12.8 percent); the percentage of households that earn $150,000–$200,000 is also lower at 2.1 percent, which is more than half the average for America (4.8 percent). Finally, the percentage of people making over $200,000 is 4.6 percent nationally versus 1.8 percent in West Virginia (almost a third). The percentages are dropping nonlinearly as you might be able to tell in comparison to just the US average, which is very bad and likely a reflection of this same idea of low educational attainment.

320 Bill "Trey" Gates, who shares my nickname for being a "third," was truly a genius, so the fact that he did not finish college is irrelevant in the usual sense. Remember there is always an exception to every rule, and he is living proof of it. The point is you can still do it, but it simply becomes harder by that much of a factor.

321 I would emphasize that college is not a guarantee for success but rather a tool to help you succeed (sometimes through connections more than knowledge). Some people do not need that tool when

your trade. For most careers you need specialized skills to succeed. And while sometimes those skills can be self-taught like with some of the great computer geniuses, for the most part, higher education is the way to go.

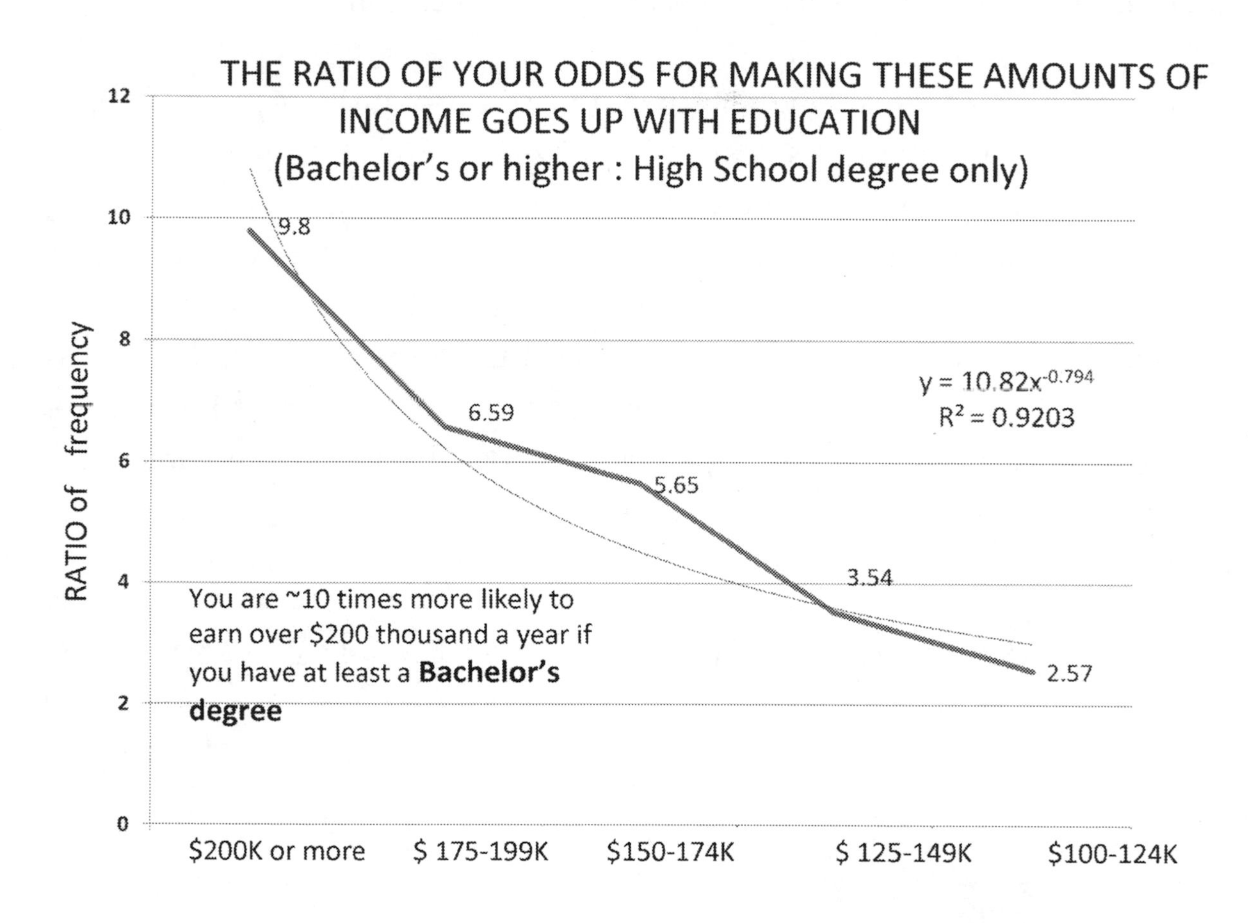

This graphic shows the increased probability of earning various threshold amounts of income (the threshold I chose was the top 20 percent of income, roughly $100,000 for household income in America) as a function of the level of education attained (the highest level for anyone in the household). If your goal were to earn $100,000 a year, then you are two and a half times more likely with somone in your household having a bachelor's degree (or higher) in America.[322] If your goal were to earn $200,000 or more, your

---

they already have other tools that will work as well or better for them, and for William "Trey" Gates this was not a required tool.

[322] Data used here is same household income data I used previously.

odds are *ten times* higher if someone in your household has a minimum of a bachelor's degree. While this is not surprising in concept, the numbers did surprise even me. This just reinforces the idea that to get more income, you increase your odds dramatically with higher education.

Over time the trend of higher education is improving somewhat, mainly because I think most of us realize by now that education pays. There is some efficiency after all in the system, although there is usually some lag period generationally that takes time to see these trends translated to our kids. If you were to examine the rates of educational attainment over the last twenty-two years, the trend is certainly one of improvement, so although the message is getting there slowly, it *is* being delivered, albeit at a different pace and with differing efficiencies within the various subgroups as I show below.

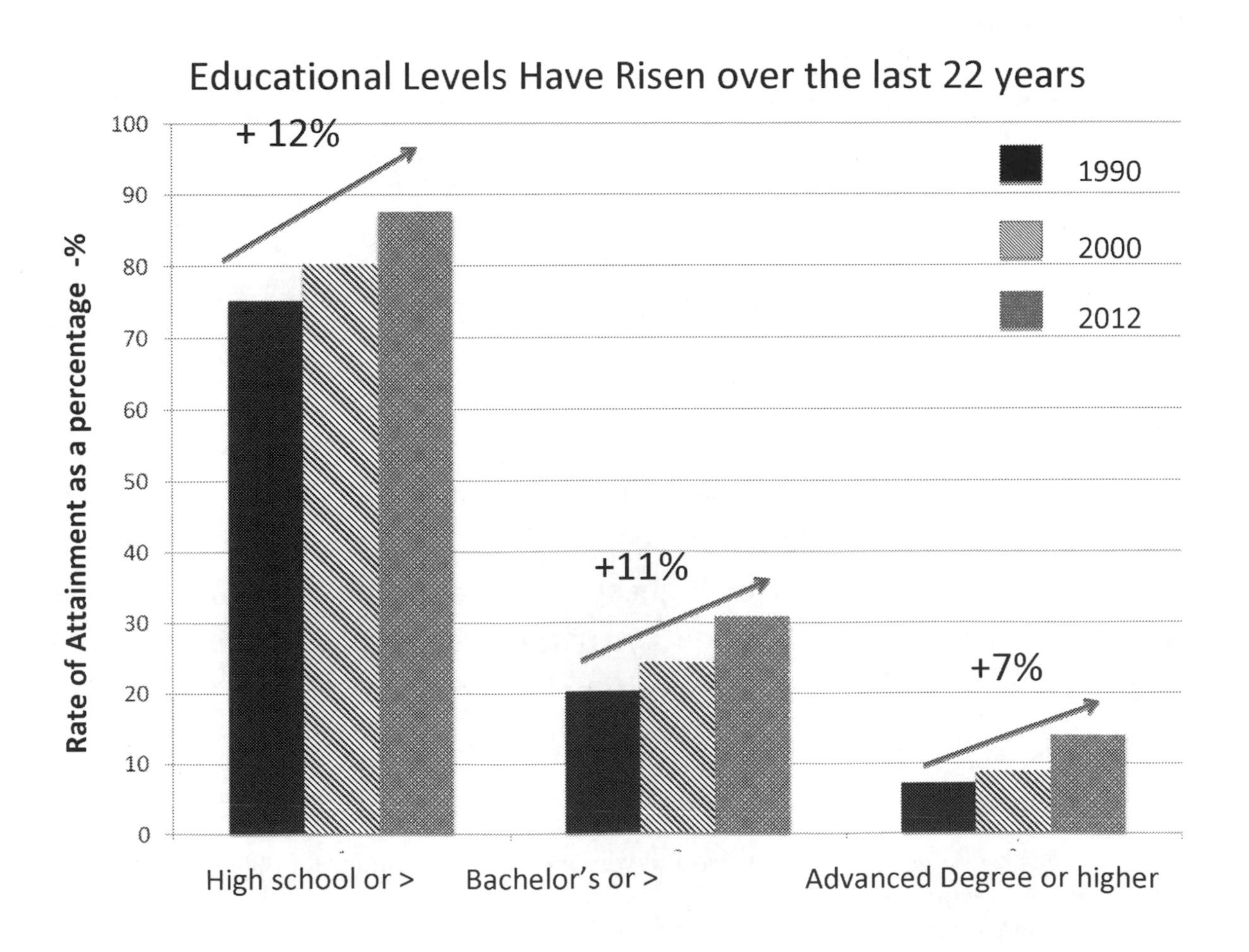

This graph shows the levels of attainment of education within America as averages for the last two generations. We are making progress and with seemingly reasonable efficiency. These are absolute rates of improvement, and if you were to look at the relative rates of improvement, it would be even more impressive. The corresponding relative rates would be 16.5 percent growth for high school graduation (over twenty-two years), a 52 percent growth for bachelor's or above, and a 94 percent increase in advanced-degree attainment.[323]

Within states and local geographies, education is even more complex compared to the national averages. Clearly certain geographies and regional cultures value and promote education more than others. I use West Virginia as an example since its results cannot be blamed on ethnic reasons[324] or on gender differences, as some of the uneducated tend to suggest (and since I lived there and experienced it firsthand). For example, where I previously lived in West Virginia consistently experiences the lowest educational-attainment rates in the nation every year, whether it is for a bachelor's degree, associate's degree, master's degree, professional degree (MD, JD, DDS, DVM, DO, OD), or a doctoral degree (PhD). While this is in part due to our initial low starting point as a blue-collar state and our long history of manual labor and the low need for college degrees, the problem is still the same—the perception modeled to our children is one of low educational achievement and low value placed culturally on degree attainment, even more so than elsewhere.[325]

---

[323] In America high school graduation rates grew from 75.2 percent (1990) to 87.6 percent (in 2012), bachelor's degrees went from 20.3 percent to 30.9 percent, and advanced degrees improved from 7.2 percent to 14 percent. US Census Bureau, "Education," *Statistical Abstract of the United States: 2012*, 153.

[324] West Virginia is 95 percent white and less than 5 percent minority, which, while skewed, is skewed in favor of nonminority population measures unless you count low educational attainment as a *minority* definition, in which case we are minorities. Our gender mix is the standard one to one.

[325] On a positive note, if you examine the rate of change by state, West Virginia is on par with the others, which suggests our kids are getting the message despite their cultural barriers. We increased our relative rate of attainment for all college or higher degrees at the same pace as the average US rate (41 percent for West Virginia versus US average of 37 percent in bachelor's from 1990–2009 and 40 percent for West Virginia versus 43 percent for US advanced degrees from 1990–2009). The

If we were to think efficiently here as a nation (or in this case as a state to try to catch up to the others), it would make sense to increase the relative rate of college graduates by making it more affordable to different socioeconomic groups, by increasing graduation rate percentages and efficiencies, and by retaining more freshmen rather than letting them drop out. This is where I believe our national focus should be if we wish to improve our overall educational efficiency within America. And these concepts do not necessarily require more money. We already spend more per capita than any other nation in the industrialized world on our education. This change simply requires a shift in the way we think about and value educational achievement as a nation. At the same time, we must maintain a high level of standards to ensure that graduates don't simply get pushed through with a paper diploma.[326] If we increase the rate of college graduates with a bachelor's or higher, the general quality of life will disproportionately increase for more of our citizens. The new norm will be college-level attainment, and the result will be a higher standard of living and an expanded middle class.

A Skewed Choice of College Majors

On a related topic, college degrees and colleges themselves are similarly related in their outcomes of popularity. When I was a kid, you went to college to mostly get a liberal arts education, and if you were lucky, your major helped you get a job. Now the majority of kids are deciding on their majors before they pick their college as a way to be more efficient in the job-match process (tools seeking jobs, as Zipf alluded to earlier), which I think is good. However, when you look at what is popular and what performs well,

---

problem is the local business culture does not support the need, so even if our youth do attain higher degrees, they move elsewhere to find gainful employment. This can be a self-perpetuating spiral of downward mobility that then adds to the creation of disparity in our education.

[326] The growing base of Internet-based online degrees concerns me. The impetus is to educate and graduate, and there may be some inherent compromise in quality in order to achieve higher graduate rates. I address the idea of balancing quality with efficiency in part 3 of this book, and somehow we need to hold these schools accountable to the same degrees as other schools; otherwise these degrees are a waste.

there is some discordance, and as shown elsewhere, more popular is not always good!

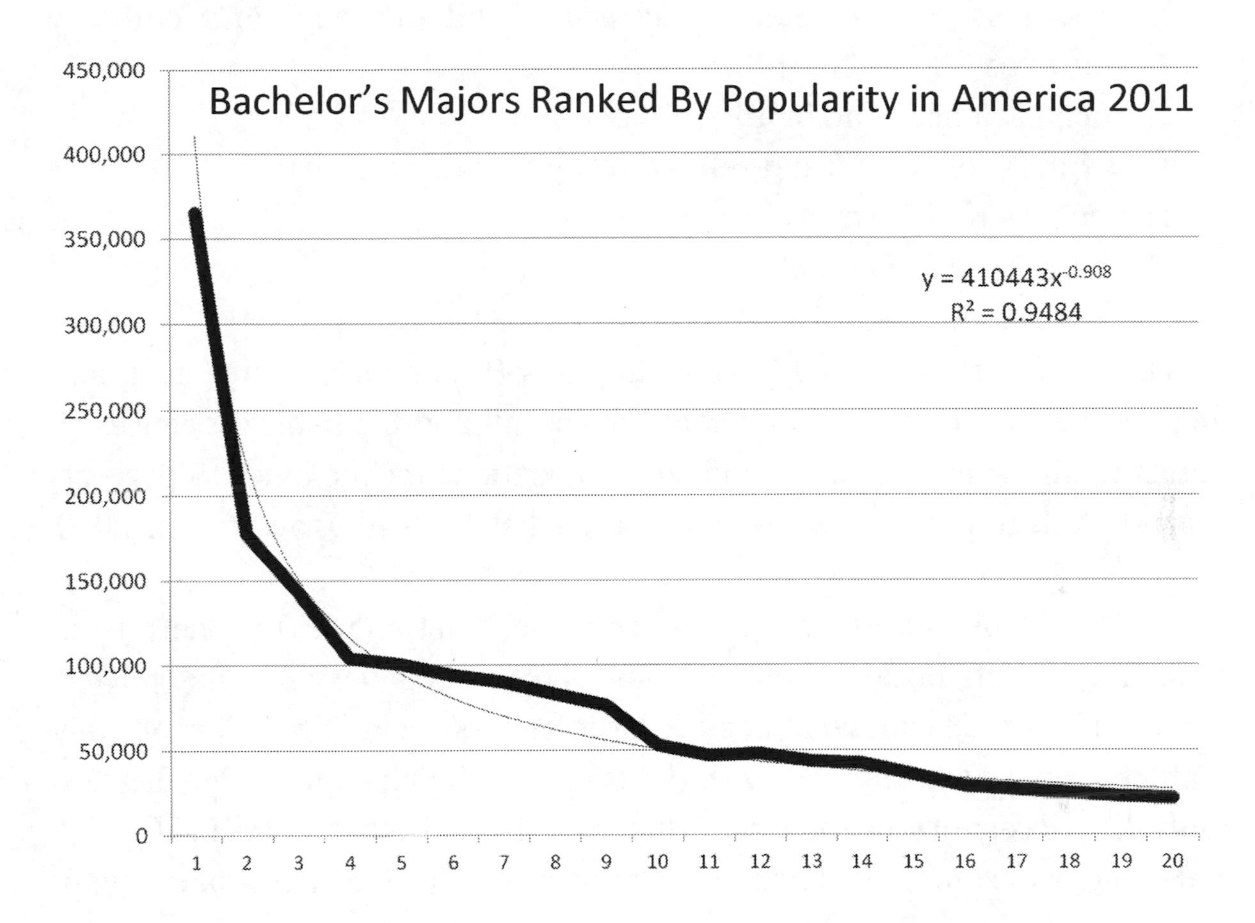

Americans tend to be very skewed toward a vital few (popular choices) majors, as seen from the statistics above (referenced majors from National Center for Education Statistics 2012). These are actually degrees conferred in 2011 using bachelor's degrees only. Popular may not always be better, and your choice should be something less popular if you are considering the idea of more for less. Tinkers, i.e., engineers, are number 9 on the above graph, and the number one major by popularity was business.[327]

[327] For a full listing, see National Center for Education Statistics, Table 322.10: Undergraduate Degree Fields, http://nces.ed.gov/programs/coe/indicator_cta.asp#info. The information is in excel format with the most up-to-date data being from 2010–2011, which is what I graphed.

For 2010, more than half of bachelor's degrees awarded were in five fields:

1. Business, management, marketing, and personal and culinary services (22 percent)
2) Social sciences and history (10 percent)
3) Health professions and related programs (8 percent)
4) Education (6 percent)
5) Psychology (6 percent)[328]

The fields of visual and performing arts (6 percent), engineering and engineering technologies (5 percent), biological and biomedical sciences (5 percent), and communication and communications technologies (5 percent) represented an additional 21 percent of all bachelor's degrees awarded in 2010.

You might be able to appreciate from the chart above that there is an imbalanced distribution of majors that seems to be a very Zipf-like power-law distribution. From what I can tell, the majors seem to be appropriately matched where the jobs are, with the exception being engineering degrees (only 4.5 percent of students chose this major in the United States in 2011[329]). This suggests some efficiency in the educational system with matching of careers with majors. It also suggests that students are appropriately majoring in the areas that give a better likelihood of getting hired or paid more.

This is what I alluded to earlier when I said students are more selective in the current generation about what they go to college for, especially when they consider what majors pay best after graduation.

Best-Paying College Majors for 2012–13 (Annual Salary)

1. Petroleum engineering ($163,000)
2. Aerospace engineering ($118,000)

[328] Ibid.

[329] This trend is relatively stable, but as a reference, in 1970–1990 the percentage of engineers as majors was over 5 percent, so the trend is slightly down, even knowing it pays.

3. Actuarial mathematics ($112,000)
4. Chemical engineering ($111,000)
5. Nuclear engineering ($107,000)
6. Electrical engineering ($106,000)
7. Computer engineering ($105,000)
8. Applied mathematics ($102,000)
9. Computer science ($100,000)
10. Statistics ($99,500)

Tinkering (engineering) pays well as you can see here, perhaps in part as a result of the disproportionately low supply of engineers.[330]

What you choose as a major and what you decide to do with your chosen major (i.e., your work type) both affect your future earnings. So while I won't tell you what to choose as a major, I do think you should strongly consider this information as part of the life equation, especially if you do not plan to get a higher degree following college. If you plan to stop at a bachelor's degree, your choice of major is where your best return on effort will be found.

For example, people working in architecture and engineering (i.e., tinkers) as a group earn $3.4 million on average over their lifetime), Computers and math ($3.2 million), management ($3.2 million), business and financial ($2.7 million), health-care practitioners and technicians ($2.6 million), sales ($2.5 million), and science ($2.5 million) all earn more than the overall bachelor's degree average of $2.4 million over a lifetime of work.[331] This would therefore translate into relatively efficient outcomes for these areas of employment if you were to do the appropriate calculations.

---

330 http://www.payscale.com/college-salary-report-2013/best-majors-for-making-money; the data is for bachelor's degree holders only with ten-plus years of experience (not starting salary).

331 Tiffany Julian, *Work-Life Earnings by Field of Degree and Occupation for People With a Bachelor's Degree: 2011 American Community Survey Briefs*, US Census Bureau, October 2012, 2–3. Data represents only bachelor's degree holders as a subset of the total population (i.e., no advanced degrees).

People who *majored* in engineering ($3.5 million lifetime income), computers and math ($3.1 million), science- and engineering-related fields ($2.6 million), business ($2.6 million), physical science ($2.6 million), or social science ($2.5 million) also earned more than the average bachelor's degree ($2.4 million). Engineering majors earned the most with $3.5 million, although it varied based on managerial roles ($4.1 million) versus service workers ($1.4 million). Notice from above, though, that less than 5 percent of majors were in this field. Education majors made the least at $1.8 million, which is sad to me, as these are the people we expect to educate our children on efficiency and who have a profound influence on our kids.[332] In fact, education majors in service jobs earned less than people whose highest attainment was a high school diploma. Yet, as I showed earlier, education majors are still in the top four most-common majors at the bachelor's level.

From an analytical perspective, thinking like a tinker that efficiency = output/input, your maximal return on the investment of your time (assuming four years and a bachelor's degree only) would be as an engineer. Output would be income multiplied by time remaining to earn income; input would be annual expense of education multiplied by time necessary to complete and get hired. Efficiency is the ratio of the two. If you combined your engineering degree with management, it would in theory be even more synergistic.[333]

Using similar logical thinking, if you pick an expensive school[334] that does not reward you with better income as an outcome, then you are being inefficient (regardless of your major). Likewise if you take longer to graduate

---

[332] Ibid. Some would argue that supply and demand drives the salaries. I personally think we should pay teachers more than we do, given the extremely important roles they have in educating our children. To me, there is no more important value than education.

[333] Synergistic means more than the sum of the individual parts. I liken it to 1 +1 = 4. It implies a disproportionate efficiency in the relation of the two variables, and management and engineering are great examples of this. Many CEOs and plant managers have engineering backgrounds, much like Pareto and Juran.

[334] There are good tools for this now that look at return on investment for each school. While Ivy League schools do result in higher income and are clearly worth the disproportionate cost, not everyone has that ability or opportunity. If, however, you are gifted, these schools seem to find a way to get you there and keep you there.

(more than four years for example), then you are also being inefficient. If you work less time or make less income, you are likewise less efficient. So all the following need to be appropriately weighted into your consideration: (1) what major, (2) what school, (3) the cost of your tuition, (4) the time you take to finish, (5) how long you plan to work after you start your career, and (6) what job or career you plan to use your degree for.

Below are the degrees that do not reward well based on supply and demand or other reasons. While this is no reason to avoid them, just realize the difficulties going into it, as my daughter does. Also remember that you can always achieve at what you do and be in the top percentages of income earners, and more importantly you can partner up with someone (and cohabit) if that is your desire as well. Most importantly, remember that income is not always the outcome that is proportional to your happiness and well-being, although it sure seems correlated with almost every other life measure that I have found.[335]

Worst-Paying College Majors for 2012 (Annual Starting Salary; Midcareer Salary[336])

1. Child and family studies ($30,300; $37,200)
2. Early childhood education ($32,200; $45,300)
3. Social work ($33,000; $46,600)
4. Athletic training ($34,800; $46,900)
5. Human development ($35,900; $48,000)
6. Special education ($33,800; $49,600)
7. Biblical studies ($35,400; $50,800)

---

[335] There is clearly a minimum level of income that is sufficient for your basic needs, and then the rest is for leisure. The minimum is relative to what everyone else around you needs. In other words, your perception of what you need varies in relation to what you see around you. There is a concept known as the Cantril Ladder that shows this in a similar way. See image 3A in the appendix. OECD, *How's Life: Measuring Well-Being* (Paris: OECD Publishing, 2011), http://dx.doi.org/10.1787/9789264121164-en. Money (e.g., GDP per capita) matters everywhere in every country up to a point in terms of happiness; then there is a point of diminishing returns regarding the life satisfaction money brings.

[336] I list starting (median) salary and midcareer salary as well to see the range here. Data from http://www.payscale.com/college-salary-report-2014/majors-that-pay-you-back.

8. Horticulture ($35,200; $50,900)
9. Exercise science ($32,600; $51,000)
10. Culinary arts ($34,800; $51,100)[337]

These are self-explanatory I think. I would point out that I married a social worker (she did well with the MRS degree, however), and my first daughter is more than likely going into theatrical arts, which also does poorly. I am totally cool with that, assuming she considers a bachelor's degree for the many other healthy, nonfinancial outcomes mentioned previously. I would point out that these choices rank in popularity highly[338] despite the awareness they do not pay as well, which perhaps suggests people don't care about income as much as they complain they do (or at least not at first).

My point here is that it is okay to choose a major in something you like as long as you realize you might struggle somewhat financially in a society that does not value what you do as much as you do (and that is a whole other subject[339]). One of these careers will be a less efficient way to earn a comfortable living, but your definition of plenty may not be income (money is only part of the equation[340]) but rather the joy you get out of what you do every day. It may be more efficient in that case for my daughter to get a roommate to share living expenses or to live at home to save money (we would welcome that to a certain degree). My point is to be aware of this data so you can make well-informed decisions from the start.

337 www.payscale.com/college-salary-report-2013.

338 Psychology was fifth most common, education was fourth, social degrees were second, visual arts sixth, and health professions third. They are all popular, but they don't pay well. As long as you are aware of this, then you will be making an informed decision and will be okay. This pay inequality probably relates nicely with supply-and-demand laws. Popular choices simply don't do well as a rule when you look at income as your outcome. Now you will be happier doing what you love to do, but you do need to put food on the table, which takes some minimal income. You can also become more efficient in your household by having two or more income earners and sharing expenses.

339 While we see the highest-paid performers glorified on TV and in film, they are the one-in-a-million achievers and not the norm.

340 A minimum is needed, however, and roughly I would estimate it is $75,000 per year currently for a household of two adults.

Work status similarly correlates with the type of degrees. People who majored in computers, mathematics, statistics, or engineering were the most likely to report working full-time and year-round and the least likely to report that they did not work at all. In contrast, fields that were classified as arts (visual arts and performing arts) or humanities (languages and literature) had less than half employed full-time, year-round (48 percent and 46 percent respectively).[341]

As you will repeatedly hear, math, science, computers, and engineering majors and careers all do well (because they are just so darn unpopular). The reason I emphasize that here is that I always tell my kids to try to learn to like math and science because it pays (disproportionately) to learn it, to like it, and even more to do something with it!

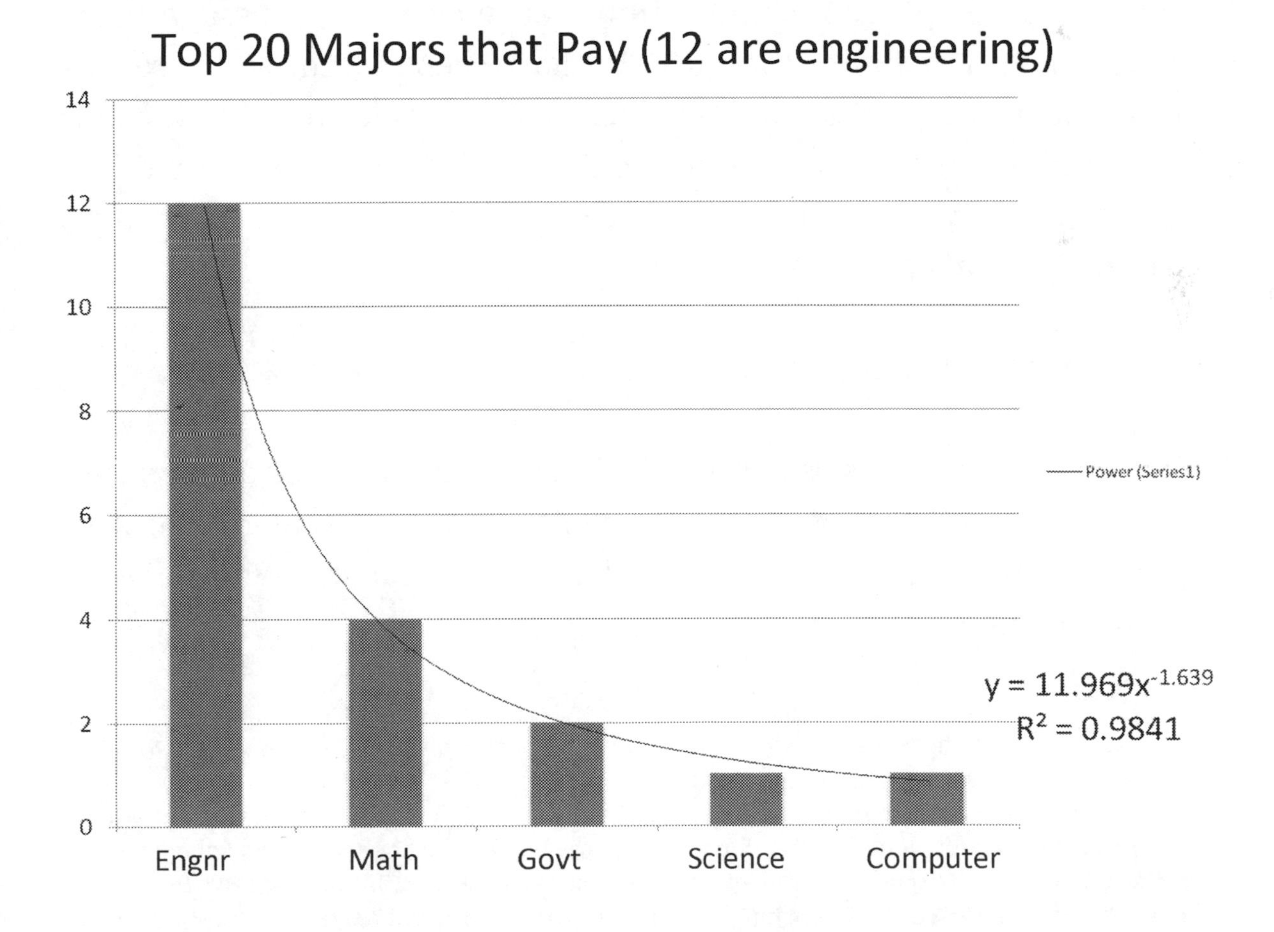

341 Camille Ryan, "Field of Degree and Earnings by Selected Employment Characteristics: 2011," *American Community Survey Briefs*, US Department of Commerce, US Census Bureau, October 2012, 1.

If you want to see the degrees that pay more, here they are graphed in rank from one to twenty and shown as a cumulative histogram. Most people currently do not choose these degrees.

Most of these top twenty high-paying college majors coincidentally are projected to grow in their respective areas through 2020 (and likely beyond).[342] Interestingly, this message is being delivered in one way or another to prospective students, at least worldwide if not domestically. According to one international report, one-third of foreign students are planning a science-related career with one in ten planning either engineering or computing. This is from a study looking at fifteen-year-olds in various countries and their expectations, not the real outcomes. The reality is that once many get there and see how difficult it really is, the numbers fall precipitously.[343] However, why is the message being conveyed elsewhere more efficiently? We have seen that the actual results in attainment are significantly less than what people "plan" (much like college graduation rates in general). In other words we have mismatched expectations between inputs and outcomes.[344] That sounds like typical human behavioral efficiency!

School Choices Matter Disproportionately

Different colleges are also skewed in their outcomes in terms of pay scales and eventual careers and so forth. Here I rank the top twenty colleges for degrees on the basis of earnings or salaries for graduates with no other variables considered.

---

[342] Brett Lockard and Michael Wolf, "Occupational Employment Projections to 2020," *Monthly Labor Review* (January 2012): 84–108.

[343] In the United States, at age fifteen 9 percent of all students (16.4 percent of American males and 2.7 percent of American females) planned to go into either engineering or computing. Poland had the highest numbers with an overall rate of 20 percent at the same age. OECD, *Education Today 2013: The OECD Perspective* (OECD Publishing, 2013), http://dx.doi.org/10.1787/edu_today-2013-en. My point here is the mismatch between what we think we want and what we actually achieve.

[344] Data is from 2010. Ibid. The United States was in the bottom eleven of thirty-seven countries in rank (we were twenty-sixth of thirty-seven total in this rank variable).

Colleges with the Highest Earnings in the United States

1. Harvey Mudd College
2. US Naval Academy
3. California Institute of Technology
4. Stevens Institute of Technology
5. Babson College
6. Princeton University
7. US Military Academy
8. Stanford University
9. Harvard University
10. Brown University
11. Massachusetts Institute of Technology
12. Colgate University
13. Yale University
14. Polytechnic Institute of New York University
15. SUNY-Maritime College
16. Cooper Union for the Advancement of Science and Art
17. Tufts University
18. Haverford College
19. Washington and Lee University
20. Lehigh University

Starting salaries from graduates of these different schools ranged from an average of $73,300 per year at Harvey Mudd (high) to $58,500 per year for Lehigh (low), and all these schools average over $110,000 per year for midcareer.[345]

To estimate the true efficiency of these programs you would need to compare the current costs (tuition and other costs) with the current salaries

[345] http://www.payscale.com/college-salary-report-2014/full-list-of-schools. It changes regularly but not that much relatively.

and then compare results. Better tools are actually now available online to do this as well, thanks to similar-minded thinkers.[346]

So if you were so inclined, you would ideally pick the best school and the best degree for an even better result. In theory the two would be not purely additive but better than either one alone. For example, if you were going into engineering, you would match yourself with an engineering college and would likely increase your results more than just with the major alone. This is why it helps to think about your major *before* you get to college.

The following simple graphic shows the powerful difference a school can make in terms of future earnings:

Average Salary and Estimated Lifetime Earnings for 44 years based on the level of Educational Attainment and for Specific Schools

Degree/School	*Avg Salary*	*Cost of Degree*	*Lifetime Earnings*	*Net Earnings*
High School Diploma only	$32,552	$0	$1.3-1.4M	$1.3-1.4M
College, *average* Bachelor's Degree 4 yrs.	$54,756	$50,000	$2.3-2.5M	$2.2-2.45M
UVA Bachelor's Public College-4 yrs.	$69,800	$195,920 (in state $50,000)	$2.8-3.0M	$2.6-2.8M ($2.5-2.75M)
WVU Bachelor's Public College-4 yrs.	$59,750	$116,176 (in state $40,000)	$2.4-2.6M	$2.3-2.4M ($2.35-2.55M)
Princeton Bachelor's Private-4 yrs.	$97,650	$216,000	$4.0-4.3M	$3.8-4.1M

346 http://www.payscale.com/college-roi/.

This graphic assumes everyone starts work at twenty-one and works to age sixty-five for an average of forty-four years. A high school graduate will incur no additional costs and will earn on average $1.3 million over his or her lifetime. A bachelor's degree from where I graduated, UVA (which was the number two public university in the United States in 2013), will earn twice that amount, and a bachelor's from a highly ranked school like Princeton (ranked number one in quality in a tie with Harvard for all schools in *US News and World Report* in 2012) will earn you three times that amount. While that may not sound like much, trust me, it is relatively speaking a lot.[347] And this graphic does not factor in your choice of major, which would further magnify the effects.

I also recommend looking into these ideas well before college begins, for example with internships and/or mentoring programs. A good friend of mine, who is an electrical engineer, spent his summers in high school interning for an engineering group in Louisiana, and as a result, he knew that would be his choice of major before matriculating to college. As a result he was able to choose a college that was better geared toward engineering rather than just an ordinary state school that emphasized liberal arts, for example. Sometimes this works out anyway if you are lucky and happen to be in the right state, but other times, the school you attend may not excel in the program you are initially interested in or the ultimate degree you decide to pursue. So take the time, make the effort, and do your homework ahead of time!

[347] I used US Census data for 2012 for high school grads versus college graduates and www.payscale.com data for Princeton, UVA, and WVU from 2012 to show specific data for schools. I averaged the salary from starting salary and midcareer salaries (twenty-two years each) to get the final lifetime average (and in each case I assumed the same forty-four years of employment for each graduate). The rankings I refer to come from 2012 and 2013 *US News and World Report* rankings of universities. Forbes and Kiplinger also had UVA as number two, just so you don't think I am biased toward my alma mater. California had five of the twenty top schools in 2013–14, which is another obvious skew for public schools of higher learning.

Why These Factors All Matter So Much

Students need jobs, and they need to repay their loans as quickly as possible if they wish to be efficient and enjoy a decent quality of life (less debt means more financial freedom). Over 50–60 percent of high school grads attempt college, but only 30 percent get bachelor's degrees (most of whom stop there). So many borrow money for college but never get to the finish line. For those people, college does not help in terms of income earnings significantly since they did not complete a degree.

Recent data also show about 50 percent unemployment rates in the graduate pools emerging from college, suggesting that a degree by itself is not enough. While it is needed, it is not *all* that is needed. Other traits and skills are required to be employable. Additionally, the average youth then carries this debt forward until it can be repaid. This adds an enormous burden to an already stressed kid and can carry over into a large part of their adult lives as well. Four out of ten students who took out loans and who are now above thirty-five years of age are still paying them back. These adults, through the power of hindsight, wish they had picked a major that allowed them a more efficient tool to repay their debts.[348]

We need to somehow teach our children the value of educational achievement and efficiency. How we do this is anyone's best guess, but our values need to change if we wish to compete in a world that is becoming flatter and more connected socioeconomically. The outcomes can change only by altering the inputs and focusing on the efficiency of our efforts and of our quality within the American educational system.

Although I focused on income as the main measure, it is simply one of the outcomes many adults value most. Money as a tool allows us to reshape the use of our time, through our utility, and is the most obvious relation to show this correlation.

---

[348] William McGuinness, "Half of Recent College Grads Work Jobs That Don't Require a Degree: Report," *Huffington Post*, January 29, 2013, http://www.huffingtonpost.com/2013/01/29/underemployed-overeducated_n_2568203.html; Tyler Kingkade, "Student Loans: New Survey Finds College Grads Carry Large Debt into Middle Age," *Huffington Post*, May 2, 2012, http://www.huffingtonpost.com/2012/05/02/student-loans-college-debt_n_1468831.html.

Education, however, correlates with more than simply income! As Mr. Jefferson[349] said, "Knowledge [education] is power!"

Higher education correlates positively with all of the following positive outcomes:

- better jobs
- better health
- less unemployment
- higher job satisfaction
- better pension (retirement plans)
- better health-insurance coverage
- less poverty
- less public assistance (welfare)
- higher likelihood of success
- more volunteerism and community service
- more active participation in government
- higher marital happiness and success
- greater life satisfaction [350]

In summary, educational attainment clearly correlates more than anything with nearly every desirable outcome and is the top priority to improve your odds for success. It takes, however, significant (i.e., not minimal) effort and requires achievement and hard work. As we often tell our own kids, "We can do hard things." But what society seems to be looking for more often is not just a diploma showing your accomplishment but another measure of your ability to succeed, and that has to do with the idea of quality. Part 3 of this book focuses on another engineer, Joseph M. Juran, and how you can achieve quality using a related principle of imbalance referred to as the 80-20 principle.

---

349 Thomas Jefferson (known as Mr. Jefferson at UVA, which he founded) could be considered a tinker (he was both an inventor and an architect). I am more simply a *thinker* but am grateful for how he profoundly shaped my education.

350 US Census Bureau, "Education," *Statistical Abstract of the United States: 2012*.

PART 3

Juran and the 80–20 Principle

"What Will I Be? Lady, Baby, Gypsy, Queen."

Joseph Moses Juran was a Romanian-born immigrant who came to America as a young boy. Like other immigrants, his father was looking for a better life for their family. Joseph Juran would later achieve the American dream by becoming an engineer in a country that disproportionately rewards quality thinkers. Juran helped establish another relevant principle along these same curvy lines of imbalance, thanks to his choice in career and his abilities to critically think (he was both a tinker and a thinker). His principle is in essence an extension of the same concepts Pareto and Zipf described, and he developed it by using the measured outcome of *quality*.

To Juran, his lifework was a search for the *underlying principles* or *universals* that, like simple algebraic formulas that he learned as an immigrant American child prodigy in math, could be used to model complex human behavior. He would succeed at finding and developing these principles, especially in business and later management.[351] Juran noted and later applied the principles of imbalance (after familiarizing himself with Pareto's lifework[352]) to describe what was initially referred to as the *law of the vital few and the trivial many* (or *the useful many*, as Juran eventually preferred to refer to it) in order to root out quality losses and improve

---

[351] John Butman, *Juran: A Lifetime of Influence* (New York: John Wiley, 1997), 12.

[352] Ibid., 48. Joseph Moses Juran learned about the Pareto principle sometime in the late 1930s to early 1940s (supposedly during a visit to General Motors) and applied it to the idea of quality management, which was his area of interest from his work at AT&T (Western Electric Company). He would later produce his *Quality Control Handbook* in 1951. Juran loved his work so much that he reportedly rarely took vacations. His job was efficiently paired with his passion, which is an important concept to learn in your life. Although he would never practice law, Juran became a lawyer during the Depression (as a buffer); he learned much about the skill of communicating and "the meaning of the language" from this additional educational experience.

reliability of industrial and consumer goods. Most importantly he added the human dimension to quality management by emphasizing the importance of education and training of high and midlevel managers in business and by learning to isolate human relations problems.[353]

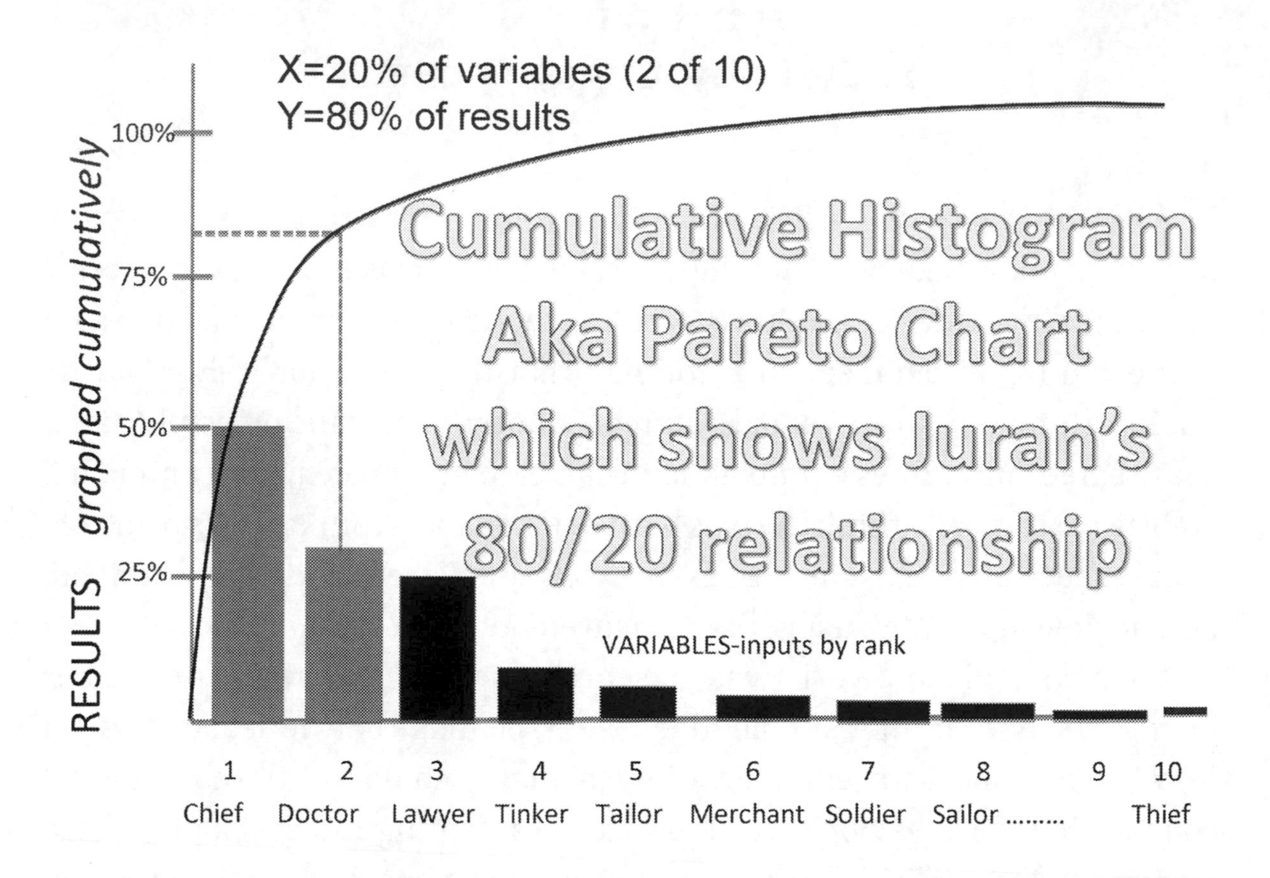

By using a graphic tool known as a cumulative histogram,[354] Juran showed the relation of the skewed Pareto distribution to fit an 80–20 relationship. This is now referred to as a Pareto chart.

Many believe Juran was the true pioneer of what we now think of as the Pareto principle of imbalance, as Juran was able to successfully apply Pareto's rule to areas of quality production and quality management. For

353 *Wikipedia*, s.v. "Joseph M. Juran," last modified August 6, 2104, http://en.wikipedia.org/wiki/Joseph_M._Juran.

354 He references mathematician Max Lorenz as well in the credit for this relationship graphically. Lorenz showed how this can be represented using a cumulative histogram in 1905 to demonstrate social inequality.

this reason, some prefer to refer to the rule as the *Juran-Pareto principle* as it implies both *quality inputs* and the nonlinearly related outcomes that follow the power-law distributions you have seen throughout this book.

As an inspector on an industrial assembly line in his early engineering career and with his eye for detail, Juran was able to analytically identify a vital few mistakes and defects that limited productivity or resulted in substandard quality. In those days, defects were thrown away and wasted, often resulting in losses of time (i.e., labor) and money.

By Juran's train of thought, it made more sense to produce something properly (whatever it was) than to make it incorrectly and then fix it or throw it away (as was the practice at Western Electric, his first career employment).[355] Twenty years later (around 1946) while working for Gillette, he applied these ideas to improve productivity by statistically identifying the faulty inputs (which turned out to be *20 percent* in the case of Gillette razors) and fixing these issues. The result was to reduce the costs of these inefficiencies by 76 percent (an example of a 20–76 relationship of imbalanced inputs to outputs, costs in this case).

This is a simple metaphor for my kids: "If you get the inputs right the first time, it will save you a lot of expense and much headache later." To quote Juran, "Doesn't a product [my translation = kid] that is built or assembled properly from the beginning generally look and perform better than one that has been reworked?"[356] To that rhetorical statement I cannot help but offer the obvious reply, "Yes, Mr. Juran! I couldn't agree more."

Juran discovered as a quality inspector in production lines that a vital few of the problems accounted for most of the bad outcomes, and the remaining contributions were trivial. This idea of sorting through the inputs and outcomes to find which variables were disproportionately responsible

---

[355] An ounce of prevention is worth a pound of cure. Preventing the problem is more efficient than dealing with it after the fact. This idea has implications to human behavior and relates to my line of work—(preventative) medicine.

[356] Butman, *Juran*, 84.

for which outcomes was very important, and to do this he used Pareto's principle of statistical analysis.[357]

For example, during the early period of World War II (before America was officially engaged in the war), Juran worked in a federally appointed office under President Roosevelt) that was designed to streamline war supplies to the Allies in a program known as the Lend-Lease effort. The idea was to help supply needed items to other countries in conflict with Germany in a way that did not directly implicate America. Juran encountered tremendous inefficiencies within the supply-demand imbalances imposed by the war. He was able to remove redundancies and streamline efficiency in the war effort; for instance, he reduced the processing time for war requisitions from *ninety days* to *fifty-three hours*. All this was through his analytical thinking and applications of statistical concepts to reduce redundancy in government bureaucracy.

Juran patriotically felt the efficiency he achieved was a "very useful effort" at the time. Years later when a reporter asked him about the most extraordinary thing that had happened in his lifetime, Juran replied, "The time I spent in World War II in the federal government program as an assistant administrator in the Lend-Lease program. It opened my eyes to some very big forces in the economy, world affairs and politics."[358] Some would argue that these efficiencies helped gear up the production capabilities of American companies so that America was able to quickly enter the foray with expedited efficiency once the Japanese[359] involved the United States directly with the bombing of Pearl Harbor.

Juran had shown that these same principles were transferrable from industry and quality control to government bureaucracy and efficiency.[360]

---

357 In a Pareto analysis, the variables are ranked from most important in rank to the least. This forms a rank distribution, which follows power-law mathematics (and usually is Zipfian with a power exponent of −1). Juran first described this analysis in his first edition of the *Quality Control Handbook*, in a caption to the "Maldistribution of Quality Losses" graph. When this is graphed as a cumulative histogram, you get the 80–20 relationship (more or less).

358 Ibid., 65.

359 Ironically the United States helped a vanquished Japan rebuild its economy, only after removing all existing elite businessmen and CEOs. This allowed a clean slate for Juran and others to enter into a devastated nation and help rebuild efficiently using his concepts of quality management.

360 Ibid., 65.

After the war, he began his career in writing, and his first book was a detailed analysis of some of the inefficiencies in the bureaucracy of government, calling for "the functions of the federal government ... [to be] performed using a minimum of personnel, time, equipment, and materials."[361] While we may not have learned this degree of federal efficiency quite yet in 2014 (especially since government is larger and seems less efficient[362]), the idea he pioneered still has sound merits.

Juran would later become a faculty member at New York University (in industrial engineering), and he soon began his real passion as an independent consultant for industry using the newly discovered methods of quality controls and statistics. During this time and through his associations with others of similar mind-set,[363] the occupation of *management* became a real entity within America and took on real status worldwide.

Juran's ideas ironically found a big following in the 1950s in a postwar culture overseas. The Japanese were recovering at a snail's pace after World War II, thanks in part to the war-related devastation of the country and in part to our political postwar efforts to limit their growth capabilities. For example, immediately after the war, most high-level executives in Japanese companies were forced to retire or quit as part of the process of penance. As most high-level managers and engineers were not able to return to work, there was an eager and desperate population to learn new business ideas (from Americans) to apply to their consumer industries that could help to rebuild the depressed Japanese economy (including auto production and electronics industries).

Juran's principles of efficiency and quality controls combined with Asian work ethic would radically transform international consumer-goods production in Japan following the war and would catapult Japanese growth

---

361 Joseph M. Juran, *Bureaucracy, A Challenge to Better Management: A Constructive Analysis of Management Effectiveness in the Federal Government* (1944).

362 There probably is a size after which efficiency is lost, and for every organization there is likely an optimal match of size and efficiency (including a country and its federal and state divisions).

363 Peter Drucker and Dr. Edwards Deming were a few key associates. Interestingly Deming developed the statistical methods for the 1940 census, which was the first to use sampling methods to poll Americans on a variety of issues. Butman, *Juran*, 78. Drucker also authored a book in 1954 entitled *The Practice of Management, A Study of the Most Important Function in American Society.*

to surpass even that of the US powerhouses. US companies would later learn from the model and eventually adopt these same quality-management principles in the 1960s and 1970s, after they had already been tried and proven for twenty years in Japan.

Juran's ideas were successfully applied within the United States by companies such as IBM to make operating systems (and software) more efficient than those of the competitor's.[364] Steve Ballmer, former Microsoft CEO, recently noted again that most of the errors from software stemmed from a few software bugs that, if fixed, would satisfy the majority of the customers: "About 20 percent of the bugs caused 80 percent of all errors, and 1 percent of bugs caused half of all errors." [365] Juran's law of the vital few has been adopted by many businesses since then in terms of the general idea and its useful applications. It is simply a repeating law of human behavior that is a powerful tool once discovered, and if companies like IBM and Apple use this law, perhaps we should too.

More recently Pareto's principle of imbalance, or the law of the vital few using Juran's terminology, has been generalized to the *rule of 80–20* (or *20–80*) due to these mathematical consistencies of imbalance. As Juran noted, when variables intimately involve human inputs and related outcomes that are desirable, the two are often interdependent with a consistent pattern. When the inputs and outputs are plotted by rank (as a cumulative histogram), they describe a 20–80 relationship, where a few of the input variables, roughly 20 percent, are disproportionately responsible for most of the related output variables, roughly 80 percent.[366]

---

[364] IBM developers recognized that 80 percent of a computer's time was spent executing about 20 percent of the operating code, so it rewrote its software to make the most-used 20 percent more user friendly and accessible, thus increasing speed and efficiency. Interestingly Benoit Mandelbrot, who worked for IBM, came across a paper from Zipf and became one of the biggest proponents of the importance of power-law math in the past few decades. Mandelbrot's PhD dissertation was on long tails, and he later developed a branch of math and science known as fractals, which has applications to medicine. He passed away recently, but his ideas will always remain a part of our knowledge base.

[365] Paula Rooney, "Microsoft's CEO: 80-20 Rule Applies to Bugs, Not Just Features," *CRN*, October 3, 2002, http://www.crn.com/news/security/18821726/microsofts-ceo-80-20-rule-applies-to-bugs-not-just-features.htm.

[366] Where this number 20–80 comes from is anyone's guess. M. O. Lorenz, another economist from Pareto's time period (1905) published a graphical representation of the distribution of wealth, which

In Juran's words, "In any series of elements to be controlled, a selected small fraction, in terms of number of elements, always accounts for a large fraction, in terms of effect."[367] Juran refers to the items that account for the bulk of any effect as "*the vital few.*"[368] These items must be dealt with on an individual basis. The rest he originally called *the trivial many,* although he later amended the description to a more euphemistic *the useful many.* These latter factors do not warrant individual attention and should be dealt with as a class.

I colloquially referred to all of these principles together by the simpler *life-isn't-fair rule* for my kids when they were younger, but mathematically the 80–20 principle better describes these consistently observed nonlinear relations. (It really isn't fair in this skewed sense, and you need to learn the rules and play by them if you want things to go reasonably well for you). Whatever you wish to call it, here is roughly what it means:

For many of life's complex events, roughly 80 percent of the cumulative effects come from 20 percent of the causes.[369]

In addition to the idea of preventing defects or problems, Juran also understood a second component was necessary to quality. Even if a producer of goods or services could achieve perfect performance with no defects and 100 percent conformance to specifications, the product or service may not be valued by society. The product may not in fact have any use or appeal to anyone. Or it may duplicate the usefulness of an already existing product. Something else was therefore needed to make the idea acceptable and useful to the consumer. For some this related to the popular idea of equating

---

is similar to the cumulative histogram that we currently use. *Journal of the American Statistical Association* (1905)

367 Butman, *Juran*, 143.

368 Ibid., 143.

369 You can either choose to emphasize the 20 percent component as I often choose, or the 80 percent part. If you want to know the vital few inputs, then you would focus on the 20 percent inputs that give rise to the majority of the outcomes; if you are an outcome-oriented thinker, you would analyze the outcomes and backtrack to figure out which vital few inputs are necessary. They are just two sides of the same coin.

quality with worth or luxury. For others, like Juran, this idea could be analyzed by a *quality characteristic*, which could be a physical or chemical property, a temperature or dimension, a pressure, or any other measurable characteristic.[370] For Juran, quality consisted of freedom from deficiencies combined with product features that met the needs of its customers and thereby provided product satisfaction. Juran simplified this into one term, *fitness for use.*

Juran further developed and applied this *natural law* of imbalance to define *quality inputs* in order to weed out inefficiencies within industry. He became what I consider the first recognized *power thinker* and built an entire career as a quality consultant using these principles. In his real-world application of Pareto's theoretical laws, he was able to identify a few critical components (about 20 percent) that were responsible for the majority (about 80 percent) of outcomes in industry (both good and bad). Indeed, he discovered a consistently disproportional link between variables relating quality to outcomes, and he meticulously applied these ideas to business and later to management.

Juran's rule fused Pareto's ideas with his own and emphasized the *quality* of the inputs, which I think is critical. Quality is an important component of achieving efficient outcomes and is disproportionately rewarded by society as well. Juran also elegantly related simple numbers (like 20 percent and 80 percent) to the skewed findings, rather than using complex math formulas (such as logarithms and power functions) or more abstract derivations (such as differential equations) like Pareto employed.

This 80-20 principle can (in my opinion) be used to relate any variables that pertain to value and human choice—for example, the vital few of the *causes* and the useful many *effects* or any human *input* and *output*. For my children I like to use the analogy of *choices* and *consequences* to keep it simpler. The choices you make (as your own vital few important inputs) have predictably skewed consequences (the outcomes you desire or that similarly you don't want) that obey these natural socioeconomic laws. A few critical choices can have profound and lasting consequences, while many

---

[370] Ibid., 86.

other choices may be useful but contribute significantly less. This is the simple essence of power thinking. This skewed relationship is a force that is very nonlinear, ubiquitous in the universe, and surprisingly predictable—like gravity.

While not many of us could spout out the formula for gravity as a universal force (or explain how and why it works as a force), it is probably the single largest active force in the universe that has influenced who, what, and where we now are (there are only a vital few of these forces in nature). You likely realized the importance of not jumping off the high branches of a tree as a young child or the relevance of not driving your car over a cliff as a teenager. While we therefore know the laws of gravity instinctually and learn to make the appropriate choices (and experience the fruitful outcomes), we cannot always recite the mechanics verbatim (nor do we care). These laws of social physics and human behavior are no different! I want you to learn to use them instinctively and not be bothered by the technicalities of the math. It sounds easy, but I assure you it is counterintuitive to how you normally think!

For example, about 20 percent of the sections in this book probably contain worthwhile messages (with 80 percent of the useful info you need to know in those few sections), and the rest (over 80 percent) is wasted space and trivial ink. Yeah, I know—mostly junk (this is somewhat depressing to me to realize this)! Furthermore, from what you read in those few vitally important sections, I would be thrilled if you were able to really get 20 percent of my intended message—most people do a lot worse! So just a few concepts will be responsible for a disproportionate amount of your take-home message! I could argue the details *more or less* mathematically, but the idea would still be the same.[371]

In your life 20 percent of what you do with your time is worthwhile and productive (and will be responsible for most of your happiness in hindsight), and 80 percent is time you have learned to waste.

---

[371] I would bet the publisher would like to know which parts are "trivial"—they could be more efficient by editing it out. If you do the math here, if 20 percent of the book is worthwhile and of that amount you remember 20 percent, 4 percent is all you really get out of this! Life is similar in this way. We keep sifting through information and ranking things in our minds by priority for later recall. Eventually all we may remember is a simple meme, like "life is nonlinear."

I therefore wish to help you *unlearn* your waste. Become efficient at what you do by learning to use your brain and focusing on quality inputs—the rest will naturally follow. In the next part of this book, I describe how to use these ideas of imbalance from Pareto, with Zipf's ideas of our efficiency with our choices, and Juran's emphasis on quality and value to teach you how to live your life more efficiently. As another great tinker, Albert Einstein, said, "Try not to become a man of success, rather try to become a man of value."[372]

Living your life more efficiently can be done repeatedly in the choices you make in your occupation (whatever it is), in your religion, where you live, or in your core values. And as Juran nobly stated and appropriately modeled in his own life, "Whatever you do, make sure it improves society."[373]

A Few Examples of the 80–20 Principle

The following are just a few examples of the principle of imbalance in our lives, as it relates to the 80–20 principle. These examples can teach you a lot about making efficient (and sometimes hard) choices.

1) In *business:*
   20 percent of your clients account for 80 percent of your sales.
   20 percent of your customers are responsible for 80 percent of your profits.
   20 percent of my doctors refer me roughly 80 percent of my patients.[374]
   20 percent of your customers cause 80 percent of your complaints.
   20 percent of the time you spend results in 80 percent of your profits.
   20 percent of your products bring in 80 percent of your sales in dollars.
   20 percent of your sales staff account for 80 percent of your sales.

---

[372] "Albert Einstein Quotes," *Famous Quotes*, http://www.1-famous-quotes.com/quotes/author/Albert/Einstein.

[373] Juran Institute, "Dr. Joseph M. Juran: Internationally Recognized as the Father of Quality," Juran Global, http://www.juran.com/our-legacy/.

[374] It was actually 21–80 percent in my analysis of my 2012 medical practice. This shows you what I mean by not being so exact but close enough to understand the gist of the relationship and how the two numbers don't have to add up to 100 percent—since it relates inputs (0–100 percent) and outputs (0–100 percent) separately.

20 percent of (quality) problems account for 80 percent of the rejects (Juran).

20 percent of your marketing efforts provide 80 percent of the results (now figure out which ones and *Just Do It*[375]).

20 percent of meeting time accounts for 80 percent of the decisions.

20 percent of my employees account for 80 percent of workplace absenteeism.

20 percent of the opportunities to make a mistake accounts for 80 percent of mistakes.

2) In *economics:*

20 percent of the population in Italy owned 80 percent of Italy's land (Pareto).

20 percent of the richest population has 80 percent of wealth of the world.

20 percent of the richest population in the United States should own 80 percent of wealth. They currently own 87 percent, so this is still arguably within the bounds of the rule.

3) In the *financial services industry,* this concept is known as profit risk, where 20 percent or fewer of a company's customers are generating positive income, while 80 percent or more are costing the company money. The standardized returns on individual stocks also follow a Pareto distribution:

20 percent of the stocks in your retirement portfolio will account for 80 percent of your returns (focus on these).

20 percent of equities from the last five years account for 80 percent of the stock market gains for the last twenty years.

4) In *crime and geography:*

20 percent of criminals account for 80 percent of the monetary value (stolen goods) of all crime.

[375] Nike's theme has become a sort of social meme, which we associate with achievement in athletics.

A few large metropolitan areas (20 percent of the United States) account for 80 percent of the American population.
A lot of small towns (80 percent of the United States) account for 20 percent of the population (Gibrat and Zipf laws).
20 percent of the states consistently account for a majority of the repeatedly bad outcomes like poverty, poor education, poor health outcomes, poor jobs, poor achievement, etc.

5) In *insurance:*
20 percent of motorists (like two of my teenagers) cause 80 percent of accidents. I have one teenager that has had two accidents before eighteen.

6) *Common individual/household facts:*
20 percent of your clothes will be worn 80 percent of the time.
20 percent of your carpets will get 80 percent of the wear.
20 percent of those who marry comprise 80 percent of divorce statistics.
20 percent of all the rooms in my house accounts for 80 percent of my time utilization.
20 percent of my friends account for 80 percent of my leisure social events.
25% of my kids have caused the majority of our household's unhappiness.
20 percent of the food I eat accounts for 80 percent of my overall diet.
20 percent of my fitness workout routine results in 80 percent of my results.

7) In *education:*
20 percent of people attain 80 percent of educational qualifications available.
15–20 percent of web pages account for 80 percent of links on the Web.
20 percent of scientific information published accounts for 80 percent of usable information.
20–30 percent of the words we use in our language account for 70–80 percent of word frequency (Zipf).

20 percent of your time spent studying will account for 80 percent of your knowledge and test results.

20 percent of this book will account for 80 percent of its helpful information.

8) In *health:*

80 percent of deaths from illnesses are caused by 20 percent of causes (cancer and cardiovascular diseases are the major players).

75–80 percent of the cancers we see are caused by a vital few choices like smoking, diet, and exercise.

80 percent of our healthy outcomes (e.g., fitness) come from a few simple inputs that we choose (diet and exercise).[376]

80 percent of viral causes of cancer are from a few pathological organisms (human papillomavirus, Epstein-Barr virus, hepatitis B and C, human T-lymphotropic virus, etc.).

80 percent of the costs of medicine come from 20 percent of the habits we consistently repeat (e.g., obesity, smoking, diet).

9) *In religion*:

80 percent of religious people fall under 20 percent of the subtypes (Christians, Muslims, Hindus, etc.).

1 religion (approximately 20 percent of the major world religions) accounts for 80 percent of America.

In the next section, I plan to show you how this same idea (80-20 principle) relates to what is popular and what is out of balance in our value systems, using religion once more as an example.

---

[376] This is one of the initial subtle inspirations for me in this book as applied to life. Timothy Ferriss's book, *The 4-Hour Body* (the four hours is roughly per month) in essence promotes this same principle of efficiency in terms of many health choices such as getting fit. You can be healthy using a few vital principles.

"How Shall I Get to Church?"

Our Quality Values (e.g., Religion) Are Imbalanced

One of the final ideas I wanted to examine for my children, that I link to Juran and his *quality* measures, is the imbalanced distribution of our values, including religion, and how our chosen values relate to these same rules that form our internal frame of reference, both as an integral part in what shapes the values of the modern family but also within American communities. Our choices in religion and other related moral values are similarly skewed just like everything else I have examined throughout this book, and why shouldn't they be?

Religion is simply another model that demonstrates the cultural preferences of human beings, and therefore it should not be any different than education, money,[377] geography, or anything else to which we attach value.[378] But I would propose religion (and our implied moral-ethical value system) is more vital than all the other inputs we have discussed. It should be ranked highly in your way of thinking if you wish to really be happy—in fact, it really should be number one.

By no coincidence, "Tinker Tailor" has origins from an ethical treatise by an Italian monk, Jacopo Da Cessole, whose desire was to teach simple moral life lessons as a way to encourage fairness in the way we treat other human beings. Originally written in Italian, the treatise was later translated in the fifteenth century into (Gaelic) English in a book called *The Game and Playe of*

[377] I would point out that even our basic money points to the need to trust in God, which has been the motto of our nation since 1956. The reverse of the one-dollar bill has an image of a pyramid (there is the hierarchy thing again) with the Eye of Providence (God) over it, with the motto "In God We Trust."

[378] Religions even circulate throughout time in the same way everything else does, based on these same ideas of circulation of the elite as proposed by Pareto (see Pareto, *The Mind and Society*, 3:1431.).

the Chesse, which by no coincidence was one of the first two books printed into English.[379] As a Dominican monk, Da Cessole used the story effectively as a tool to teach lessons about human ethics and to espouse Christian values within society by linking honorable values to the various occupations[380] using chess players; for him it was a way to also teach the then-governing elite the various roles and responsibilities the "government"[381] had to its members, and vice versa.

In the original story, the ruling elite (a king) was an evil monarchist who ruled with fear and oppression. He was then coerced into learning the game of chess by observing a Greek philosopher who cleverly played the game with men of learning in order to craftily teach the king the virtues of life. The king eventually learned the game through observation, and in teaching the king, the philosopher educated the monarch (and the ruling elite) about how to properly govern the people and live a virtuous life:

> Than the philosopher began to teche hym and to shewe hym the maner of the table of the chesse borde and the chesse meyne

379 It is debated, but it may have been the second book printed into English. The first may have been *The History of Troy,* called *Recuyell of the Historyes of Troye* (also from Caxton).

380 The various occupations at that time were: (1) laborers and workmen; (2) smiths and forgers, carpenters, (*tinkers*), and makers of money; (3) *tailors,* drapers, cloth makers, notaries; (4) *merchants,* bankers, and *chiefs*; (5) physicians (*doctors*), spicers and apothecaries, masters of law (*lawyers*), logicians, scientists, and mathematicians; (6) taverners, hostelers, and food/wine vendors; (7) guards and keepers of cities (also *chiefs*); and (8) ribald, couriers, and gamblers and messengers. William Caxton, *The Game and Playe of Chesse,* ed. Jenny Adams (Kalamazoo, MI: Medieval Institute Publications, 2009), http://d.lib.rochester.edu/teams/publication/adams-caxton-game-and-playe-of-the-chesse.

381 The original governing elite was an aristocratic king, and the members were the poor subjects. The original treatise from Da Cessole was entitled *Liber de moribus hominum et officiis nobilium sive super ludo scacchorum* (*Book of the Customs of Men and the Duties of Nobles or the Book of Chess*) and was written well before this in the 1200–1300s. Da Cessole was an avid chess player and a monk, and he metaphorically linked the idea of morals to the conditions of men, with their various ranks represented by differing chess pieces. By creating a hierarchy of players (by way of their professions) from pawns (which included most of us common folk) to the more powerful players like the king, queen, and bishops, he educated the then-ruling elite about ideas of morality of governmental rule, the role of all the various classes within society, and how to lead a virtuous life. He related the various "Pawns of life" (the common folk) and the ruling elite (the king and queen and bishops) to their implied virtues and behavior.

> And also the maners and condicions of a kynge of the nobles and of the comun peple and of theyr offices and how they shold be touchid and drawen. And how he shold amende hymself & become virtuous.

And the philosopher tells the king,

> My ryght dere lord and kynge
> the grettest and most thinge that I desire is that thou haue in thy self a gloryous and vertuous lyf And that may I not see but yf thou be endoctrined and well manerd and that had so mayst thou be belouyd of thy peple Thus than I desire y't thou haue other gouernement than thou hast had
> And that thou haue upon thy self first seygnorye and maistrye suche as thou hast upon other by force and not by right Certaynly hit is not ryght that a man be mayster ouer other and comandour whan he can not rewle ner may rewle himself and that his vertues domyne aboue his vices.[382]

The book then goes on to describe further (in Gaelic English) the traits of moral and ethical behavior for each of the players, essentially teaching the core values of human behavior by analogy to each player of the game. Each chapter relates a player with an occupation (e.g., tailors, merchants, masters of law, and physicians) and its moral responsibilities. Most of the players are represented by pawns, except for the king and his governing elite (the queen, bishops, knights, and rooks). It makes for really interesting reading and can mostly be deciphered phonetically.

In the end, the "kynge" learns to make the correct choices for his people:

> And by this maner hit happend that the kynge that to fore tyme had ben vicyous and disordynate in his liuyng was made Iuste and virtuous, debonayre, gracious and full of vertues vnto alle peple

[382] Ibid.

> And a man that lyuyth in this world without vertues liueth not as a man but as a beste.[383]

Human existence, hence morality and the human condition, is on a larger scale very much the same game of competition and cooperation, involving a similar strategy as the game of chess. Therefore the intent of the nursery rhyme as it originated (by way of a Christian educator) to teach the roles of the monarchy (queen and king) and the church (bishop) and the citizens (pawns) to each other, emphasizing moral and ethical values we all share, is entirely appropriate.

The evolution of this elaborate moral and ethics treatise into a simple and condensed modern nursery rhyme[384] is another example of how knowledge can be oversimplified with the power of time and modified to suit the circumstances of the various geographies and time periods. More relevantly, the original intent is often lost as time and people modify it to reflect their changing values. We prefer simple, remember? Most religions, by no coincidence, have a vital few simple moral rules worth remembering as repeating laws (e.g., the Ten Commandments[385] in the Old Testament—one for each finger)[386] that help make it simple and easier to remember.[387]

---

[383] Ibid.

[384] While I am no expert on this evolution of an idea, I could argue that someone somewhere randomly made the link between the rhyme and the old story without a true natural evolution, and the idea stuck. It really does not matter, since the rhyme has all of the elements I am after. But if the link is indeed true, it is interesting and would seem to make sense and fit the natural evolution of religion and many other ideas with time.

[385] The Ten Commandments replaced hundreds of existing rules and laws at the time, which was appealing to the masses. One of the laws is to honor your mother and father (and grandmother in the rhyme's case).

[386] Newton said, "In the absence of any other proof, the thumb alone would convince me of God's existence." *Wikiquote*, s.v. "Isaac Newton," last modified September 11, 2014. http://en.wikiquote.org/wiki/Isaac_Newton. He conceived of the world as not being governed by an interventionist God but as crafted by a God that designed along rational and universal principles. To Newton, God was an essential part to the nature and absoluteness of space, and I would add time. These principles were available for all people to discover, allowing man to pursue his own aims fruitfully in this life, not the next, and to perfect himself with his own rational powers and choices.

[387] The KISS principle is all about efficiency: keep it simple, stupid! If you keep it as simple as possible, everybody can learn. If not, as my kids sometimes hear me say: "There is no cure for stupid!" I would suggest here that even though we know the rules or commandments, we are inefficient at following them.

God knows we all like simple! And while religious differences seem to persist and even grow with time, the basic tenets are fundamentally what people need in a society that struggles with the dichotomous ideas I have shown throughout this book.

Whether it is rich or poor, educated or not, the moral struggle we all face is often good versus evil, love versus hate, doubt versus faith—and religion is one of the most efficient ways to keep humans morally balanced (I should say *less imbalanced*). Unfortunately, we are losing some internal efficiency in this regard within American culture as it pertains to religion as a choice, and I don't think it is entirely attributable to science or technology, since I am a firm believer in both.

Globally, a *vital* few religions naturally account for the majority of what we see and experience throughout world cultures. Christianity, Islam, Hinduism, and Buddhism represent the top four religions (these would be Joseph Moses[388] Juran's vital few in regard to religions) by rank order. These four religions account for about 75–80 percent of all the world's spiritual believers.

Much of this is undeniably determined by where you live as well as population demographics that are interdependent and culturally driven. And remember what I said earlier about how we cannot pick our parents? Your religion is disproportionately affected by where you were born and where you live (remember the idea of propinquity?), and all this relates more to your genetic donors (i.e., mom and dad) and less to your own inherent choices, unless you can learn to think about life independently and critically.

---

388 The first and middle name references for Juran should be obvious to Judeo-Christians. The Ten Commandments are also known as the Laws of Moses.

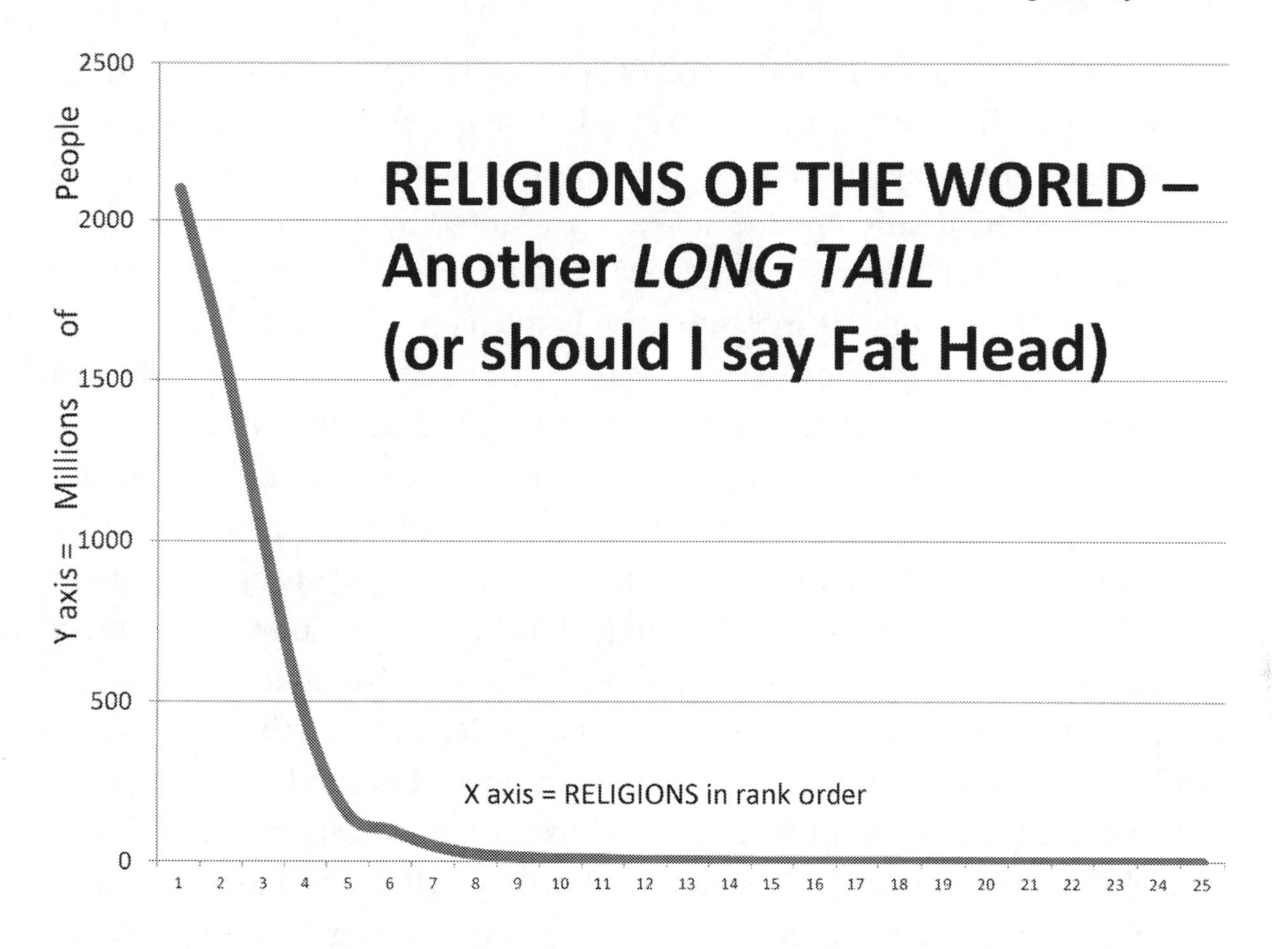

This image shows religious efficiency or religious popularity: 20 percent of the world's religions account for the majority of what shapes our beliefs in world culture.[389]

Could one religious group be truly right and the others wrong? Does anyone really know what God and the universe are all about, except God? Could we all be wrong?[390] Is what is popular by one account not correct by

---

[389] The data I used to graph this is from *Wikipedia*, s.v. "Religions of the World," last modified August 17, 2014, http://en.wikipedia.org/wiki/List_of_religions_and_spiritual_traditions. Twenty percent (four) implies there would be roughly twenty total religions of significance, and as you can see on the graph, this is accurate.

[390] Pascal's wager is a form of logic I think makes sense even to simple thinkers. He argued that humans all bet with their lives in believing that God exists or does not exist. Given the possibility that God actually does exist and assuming the infinite gain or loss associated with belief in God or with unbelief, a rational person should live as though God exists and seek to believe in God. If God does not actually exist, such a person will have only a finite loss (some pleasures, luxury, etc.). *Wikipedia*, s.v. "Pascal's Wager," last modified September 9, 2014, http://en.wikipedia.org/wiki/Pascal's_Wager.

another's?[391] Pareto himself argued, "Everybody is firmly convinced that his religion (morality, law) is the true type. But he has no means of imparting his conviction to anyone else. He cannot appeal to experience in general nor to that special kind of experience represented by logical argument. In a dispute between two chemists there is a judge: experience. In a dispute between a Moslem and a Christian, who is the judge? Nobody."[392]

We all adamantly adhere to our sentiments and beliefs. Pareto referred to sentiments and beliefs as *residues* or *nonlogica*l reasoning that persists in our culture. Science[393] follows *logical* reasoning as evidenced by universal observations that become law over time. In reality, and as suggested by Pareto, the residues are the basic underlying ideas (in this case common moral beliefs) that form the *derivations* (e.g., different religions) often with a unified idea or principle. For example, as Pareto said, "A Chinese, a Moslem, a Calvinist, a Catholic, a Kantian, a Hegelian, a Materialist all refrain from stealing; but each gives a different explanation for his conduct."[394]

If we wish to coexist peacefully in an expanding world community, the underlying common moral principles become what are most relevant. Rather than finding the unique differences in the authority of their origin and their frame of reference, we should look for the unifying principles in the different world cultures. Similarly within cultures like America, there is continual divisionism even within Judeo-Christian beliefs, but at the heart are basic underlying residues or tenets that are common core beliefs to most of us, regardless of how we choose to classify, rank, or otherwise

---

If you were a game theorist, it would make sense for you to believe, whether you had proof or not, of a higher being (this could be the definition of faith—belief without proof).

[391] I could argue that certain clusters of people are more religiously efficient; they are better able to instill the concepts into their people for whatever reasons.

[392] Pareto did not like things that could not be proven, and by definition faith is not provable. He preferred things that were logical and not based on sentiment (which he called residues), though he recognized how prevalent sentiment was in our behavior.

[393] Faith and science are two different ideas. You can never prove faith. Even Pareto commented on this in his treatise, so I will not try to prove the existence of God to the reader, lest it become no longer faith. Pareto wisely said, "Faith and science have nothing in common, and a faith can contain neither more nor less of science." Pareto, *The Mind and Society*, vol I, 13 (footnote)

[394] Pareto, "The Logic of Sentiment," *The Mind and Society*, volume III: 897

categorize ourselves. We should seek those beliefs that bind us together rather than those that separate us from others and from God.

In a way, therefore, "You can't always pick your God." I don't mean to sound flippant when I say this, but I do mean to suggest that profound influences—where we are born, who our parents are, what beliefs they adhere to, what practices our local community follows, and so on (a sort of religious propinquity)—shape even our basic beliefs in religion. I suppose I could argue that we have individual *choice* through *free will*, and we could in theory by exposure to other religions change our perspective and our view on religion with the opportunity, but you and I know that even with popular evangelism this is very unlikely in certain parts of the world, particularly in countries that do not advocate free will.[395]

Nevertheless, the hope of most religions is to expand their strongholds of popular followers, and this is part of most religious strategic plans and the idea of evangelism. If religions don't continue to grow, they will perish like the older pagan religions. But despite what we call religion, some common ideas transcend time and space and are similar within all religions, mostly having to do with treating others lovingly and worshiping one god.[396] The challenge is that most of us believe what we believe is right and the only way, and we naturally want to share those beliefs with others.

But in all honesty religion is an evolving idea, as is everything, and the more-efficient religions perform better and grow exponentially or by power laws—either based on popularity through prevailing culture, through truth, through divine influence, or through better human modeling and teachings. I often use my children as an analogy. I can teach them all that I am able and tell them what they should do, but the reality is they need to experience life

[395] This is one of the reasons perhaps for atheism and other growing disparity within religion. The more we allow free will within a geography, the more there will be a natural tendency toward independent thinking and ideas, which, although they may share a common core belief, will become more separated on trivial ideas (like Baptist versus Methodist and so on). It is an evolutionary concept of religious ideas.

[396] Pareto references Gousset in *Theologie dogmatique*, vol. 1, 325. "All races of men have preserved a more or less distinct conception of the oneness of God." Pareto, *The Mind and Society*, volume III: 928. This is one of the basic tenets of Christianity as well: Love the Lord God above all others. Love thy neighbor as thyself.

for themselves. The best I can do is to model important religious values to them, and this is the way I choose to practice my Christian faith to others as well. I can convert others to Christianity only through becoming more Christ-like in my own human behavior.

Some religions often seem so dogmatic in their teachings and likewise fixed in their ideologies that, unless you are a perfect follower, you will constantly feel some tension with other believers (and even with the religion itself). Religions that persist throughout time seem to be more adaptive and in tune with the evolution of societal values. For example, in my opinion, Orthodox Catholicism has not adapted efficiently over the last few decades to match the population's changes and will continue to decline unless it adapts more dynamically.[397] This is especially evident in the use of religion as a tool to condemn other societal outcomes, such as materialism and other things we value. There is a reason why we don't recommend discussing religions socially when we convene as groups, even among Christians.[398] Other religions efficiently prosper among the poor and oppressed and can be used as radical tools to encourage martyrdom as a way to reach salvation among those who are too immature or uneducated to think critically about their lives. Either way, religion can be invisibly used as a tool to accomplish something we want, as Zipf alluded to in his tools-seeking-jobs analogy.

Most of us tend to follow the easy path of what we already know from childhood into adulthood (it is the least effort, after all), so the idea of religious mobility (choosing a religion) becomes more and more challenging as we age. Religion tends to sort out clusters of people of similar beliefs, and if these geographic boundaries can be respected, we rarely have issues. Once we violate these boundaries and try to forcefully impose our beliefs on others, though, we experience issues with conflict.

---

[397] One example that comes to mind is the Catholic Church's stance on birth control. While I don't mean to suggest radical belief changes, I am concerned that the rigid ideologies limit new growth and sometimes create breakaway factions in religions.

[398] Even Christians are divided and skewed in their beliefs about the various different interpretations of the Bible and God and heaven and man's role on earth. Division and dichotomization is where a lot of this in fact starts, so you might imagine it can only get worse with time, unless we learn to look for unifying principles rather than dividing ones.

My own personal observation with *popular culture* is that modern generations often veer away from their religious faith as young adults only to rediscover religion (and even examine others) in later years when there is something they need or lack in their lives. For many like myself, there is a temporal development in our faith that mirrors our development as mature human beings. This rediscovery often happens through marriage and coincident with family, when religion takes on more relevance.[399] It sometimes occurs when we lose something of value, such as a loved one, or when we come face-to-face with a personal struggle, such as a life-threatening disease. For whatever rhyme or reason, religion resurfaces for many as a potential to reconnect permanently with their spiritual selves.

Of course, one of the ideas behind spreading the word of religion is that people have the choice and free will to change their personal inputs (i.e., their religion) based on what is in agreement with their own personal lives, but I would imagine changing religion is a difficult matter for anyone. Your religion is already determined in large part by where you live, your circumstances, and how you were raised in your frame of reference. Once again, it would seem that life is imbalanced and seemingly unfair from the start in this regard.[400]

It should therefore not come as a surprise to the learned power thinker that some of these same imbalances and intolerances within religion cause predictable social conflicts, including unrest and wars throughout the world's cyclical history. People are passionate about their core beliefs, whether it be religion, family beliefs, morals, or otherwise. For example, I am a firm believer in the life of Jesus Christ, but others choose to follow different beliefs or different prophets. Conflicts stem from predictable imbalances in ideology that can lead to resentment, and ultimately these imbalanced inputs can result in negative outcomes, such as in the case of war (or in the modern world, terrorism, which is a more popular and efficient tool with radical groups that use religious ideology as a tool for rationalizing their motives). We have to be careful not to always associate religion with radical

---

[399] This has been the case for me, and I have rediscovered a renewed interest in religion as I've gotten older and as I seek something greater than my own purpose and myself.

[400] For me it was not so much changing my religion but deepening my connection with it and personalizing my relationship with Jesus Christ through prayer.

beliefs of various groups, regardless of the denomination, but the reality is that humans tend to categorize people, and this often leads to stereotyping. I anticipate this will change as it has done in the past often and conflicts will resume a religious undertone as they have in the past.

I am not certain whether it is the theoretical differences themselves or rather the inability of humans to tolerate religious differences that has riddled human history with wars, but I'm sure the ideas are related. To no surprise (religious-based) wars follow power-law distributions.[401] I am pleased to see our governing elite has made extra efforts to separate religious ideology from politics, but there is no doubt that these issues are closely intertwined with one another based on the principle of (religious) imbalance in our general human behavior.

Throughout human history, there have been a few large wars resulting in many deaths and a lot of small conflicts with fewer deaths.[402] This gets into a depressing aspect of the nature of war, the accelerants of technology, and the ideas of self-organizing criticality, and critical thresholds for religious based populations which is all quite scary when you realize wars are predicted on the probability basis of these same power laws. I would, in fact, suggest that these events (e.g., religious wars) are as probable as earthquakes (and as predictably destructive). Let your government worry about this for now, but be aware that this is a real phenomenon and can be modeled by these same repeatedly predicted imbalances.[403]

If you were interested in history and cared, you might wonder how many wars throughout our history had underlying religious motives? I can come up with a few examples from past and current events, though I am no expert on history or religion: (1) the Crusades, (2) the Muslim conquests, (3) the Reconquista, (4) the French Wars of Religion, (5) the Thirty Years' War, (6) the Nigerian Civil War, (7) the Palestinian-Israeli conflict, 8) the

---

[401] See image 6A in the appendix.

[402] More specifically, the casualty rates of all wars follow a power law, as do the casualty rates within individual wars. See image 7A in the appendix. Terrorist attacks likewise follow similar power laws. Different cultures have no impact on these mathematical patterns.

[403] Game theory is a branch of mathematics that attempts to do this. By no coincidence it developed out of the strategy and mathematical probability of *chess.*

Indo-Pakistani conflicts, (9) the Second Sudanese Civil War, and (10) the Lebanese Civil War.

And how many other "holy" wars have resulted in the genocide of religious practitioners of various faiths? As I mentioned, I am no theological historian, so I am not sure if more wars have been based on differences in political ideology or theology, but either way imbalances in the way we "believe" and what we consider true "faith" have resulted in conflict that obeys the same behavioral outcomes.

Religious values have had disproportionate affects on most cultures worldwide, not only in negatively skewed ways with wars but also with scientific advancements, including medicine, which is close to the heart of my own interests. Much scientific advancement of the past millennia has been met with significant resistance from the church.[404] But despite the cause of the retardants of scientific progress, the true nature of the universe ultimately becomes manifest as a result of these power-law forces (I might even suggest these forces create a sort of preplanned mechanical efficiency).

Fortunately with modern times and our free will promoting country, we enjoy more freedoms in these issues of tolerance (for example freedom of religion and freedom of speech), but *Americans are losing both our religion and our science* (as I alluded earlier in educational attainment and less people choosing the appropriate STEM majors). This growing apathy for both concerns me greatly. Not only have we de-emphasized science (and math) within our education system, we have also removed prayer from it as well, so as not to offend the religious minorities and agnostics/atheists.[405]

---

[404] Like me, Galileo was also a man of both religion and science. The Catholic Church tried him for his heretic scientific beliefs of heliocentricity, that the sun was the center of the universe. For the Catholic Church at that time, the moral (religious) and scientific teaching was that the earth was the center of the entire universe (based on Aristotle's ideas of everything being relative to us, which in part was accurate). For Galileo to think and express something contrary to the church's beliefs was not only unpopular but also considered heresy, and he risked being killed for his beliefs, even though they were scientifically valid. It is hard for us to imagine the power of the church at that time, but it was part of the governing elite back then as it is now (though I think less so now). Instead of being executed, he was confined to house arrest for the remainder of his life, and thanks to that choice, we now have many ideas that originated from his ability to think and write in the peace and quiet of his home.

[405] I am surprised we have not removed God from our monetary basis as well (In God We Trust), although surely that is next.

Within America, for example, our *choices* in religious affiliations follow similar patterns of unequal and very skewed distributions, much in the same way as educational attainment (and income). A vital few religions account for most of believers in America, and many minor religions are less popular. While no choice of religion is perhaps inherently correct, the reality is that certain religions are more efficiently spread within the US population based on our culture and our historical starting point. Interestingly, a lot of America's religious outcomes originate from our Italian friends, the Romans (and the Dominican monks like Da Cessole) and the Catholic Church and their decision to embrace Christianity over prevailing pagan ideas thousands of years ago. Christianity also survives based on what I firmly believe are inherently great values, which are truth, unconditional love, acceptance, and forgiveness.

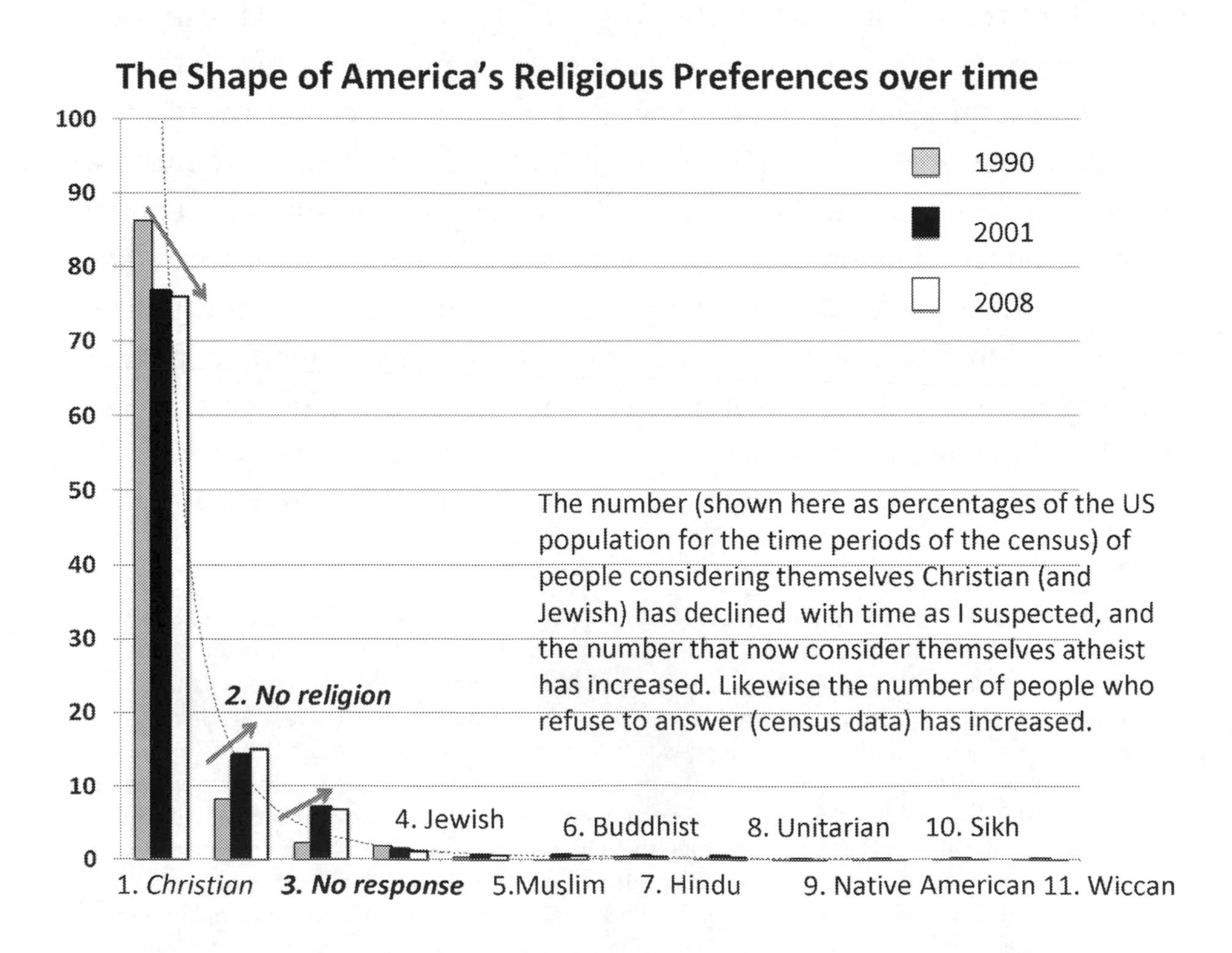

This is a graph that demonstrates the natural imbalance of religion within America and more relevantly the change of religions in America over time. We are *disproportionately Christian* in our country as you might guess (as a category this includes Catholics, Baptists, Protestants, Methodists, Lutherans, etc.) with roughly 75–80 percent of America claiming a Christian-based religious value system.[406] The second most popular choice (by self-reporting) is atheism, and third most popular is to have no choice (agnostic). The next most common religions in rank within America after (1) Christianity are (2) Judaism, (3) Islam, (4) Buddhism, and (5) Hinduism.[407]

Currently, some form of religion is claimed by just over 78 percent of the population in America, while 22 percent of Americans are either atheists or agnostic. This compares to two decades ago, when the numbers were 89.5 percent and 10.5 percent.[408] This is where what is becoming more popular is not good. Religion (and this includes the more general idea of spirituality as well) has therefore declined as a core value within our country and continues to decline with time as the graph shows. We simply choose not to think about religion, because it is easier not to and involves less effort, or we choose to not believe in God for cultural reasons or otherwise. Either way, *the growing apathy* in our nation toward not just religion but our implied moral values[409] in general is quite worrisome.

What was not surprising to me from this analysis was the growing number of non-Christians, which is perhaps a result of our growing international culture and how it has changed over the last two decades to reflect our principles of freedom of religion. What was surprising and more concerning to me was the growth in the number of *nonreligious* beliefs. This reflects a changing American culture that is starting to perhaps become less reliant on religious (and hence, by inference, moral and ethical) values.

---

406 US Census Bureau, Table 75: Self-Described Religious Identification of Adult Population: 1990, 2001, and 2008, *Statistical Abstract of the United States: 2012*.

407 See table 2 in the appendix.

408 Ibid.

409 You can of course be an upstanding moral citizen and be atheist (we have a few friends like this). But in general, I do believe there is a strong correlation with good wholesome values and religions, particularly as they pertain to altruism.

While both Christianity and Judaism have declined significantly and a few minor religions have increased somewhat (e.g., Muslims have doubled, which is explained perhaps by more immigrants), the more concerning change is *the increase in the atheist and agnostic percentage* of the American population. This number has *doubled* in the last two decades to represent *over 20 percent* of the US population!

While that statistic may not seem significant to you, I think it is prototypical of (and linked) to some of the other growing imbalances that I've discussed (including American education, health disparity, career outcomes, marriage and divorce, and general happiness). These imbalances are likely all common symptoms of a growing viral infection, in this case one of growing apathy in our core moral values (our *principles*, to borrow a word I have utilized extensively in this book), which then contribute to income disparity, poor health outcomes, diminished educational achievement, lower marital success, and generally declining happiness and well-being. Ouch! No wonder our country is depressed and taking so many antidepressants and pain pills![410] And we don't want to go to church so we don't have to feel guilty for it!

While I suggested in earlier chapters that an efficient way to increase some of our desirable outcomes would be to increase our educational and marital efficiencies, I also believe we need to reshape our spiritual and moral values, with religion as a vital part of those *quality inputs*. Whether Christianity is the choice, or any other religion that promotes similar healthy ethical values is chosen, I think the basic idea is the same. America needs to be collectively grounded in something that promotes spiritually healthy choices with our free will as espoused by the Dominican monk Jacopo Da Cessole in his treatise from the fifteenth century (that has now been abbreviated to the children's nursery rhyme "Tinker Tailor" and has lost much of its intended meaning). We need more common values that teach us to appreciate what we

---

[410] Image 2A in the appendix shows the disproportionate amounts of prescriptions we doctors write in the United States. The top twenty prescriptions account for 70 percent of all medications prescribed in America (number one is pain meds, and number three is antidepressants). We all want an easy fix for everything that ails us, as predicted nicely by Zipf's principle of least effort. Unfortunately, I think our willingness to overprescribe has contributed to a society of growing addiction to these potent medications. These drugs give more pleasure and efficiently target pain and dopamine pleasure neuro-receptors in our brain more than anything we have ever seen or witnessed in human history.

have collectively as being more than enough, and we need to learn to expect (and need) less[411] unless we change the critically necessary inputs.

I previously believed that religion was a decaying part of American society for various reasons—harsh church dogma, divisionism within Christianity, and advancements in science, to name only a vital few. I also thought in part it had to do with our society valuing money more than our intimacy with God.[412] After all, why is it so hard for so many of us to tithe as commanded in the Bible?

What I have discovered is that, like everything else, religion involves simple human behavior (and rationalizations) as outlined by the same *principles of imbalance* (of Pareto, Zipf, and Juran). The choice to adopt religion actively into our lives relates to socioeconomic principles of imbalance, involves continued work and effort[413] (and God knows we like less effort), and requires an obvious inherent value to us for it to stay the test of time. But I can assure you the quality of the outcome is way more disproportional than anything else we can accomplish here on earth if we get this input right before the end of our present time curve.[414]

Some people believe science tries to discredit religion and religion often struggles with the implications of scientific discoveries. While you may think religious doctrines exist to simply brainwash[415] the common masses,

---

[411] In the ideal world, we need only what God provides us—no more and no less. We therefore honor Him by using His gifts appropriately and generously to love Him above all and help our fellow man when we are able.

[412] In the New Testament money is mentioned more times than heaven and hell combined. In the context of the entire Bible, over 1,300 verses deal with the idea of money (and related finances over 2,000 times), so even God realized how highly we rank it. Compare this to 500 verses that mention prayer and you can see how disproportionately He knows it weighs on our minds.

[413] To be truly connected to God requires daily devotionals including daily reading of the word of God and/or private prayer. To achieve this on a daily basis is something we all struggle with, and it took me forty-eight years to master this quality effort.

[414] There is clearly a temporal evolution in the mind of an individual's spiritual development. I cannot expect my children to think as mature Christians when I am hardly mature in my spiritually in my forties (which means I have not perfectly modeled Christianity to them early on). It takes daily dedication with reading of the Bible, as well as intimate prayer with God regularly. If we don't get this ever, as some humans struggle with, then I believe we repeat life again (in another bodily form) until the lessons we need to learn are learned.

[415] I have to admit that most humans do need to be brainwashed to a degree until they learn to think for themselves and to adopt healthy habits into their daily lives that lead to happier outcomes that

as some believe,[416] I cannot help but become more spiritual and more connected with God the more I learn through science about the amazing coincidences[417] within the greater universe. I witness miracles through faith every day as I pray for the healing and forgiveness of my patients, and I see acts of godliness every time I look hard enough at the world around me for what is beautiful in the people around me and in the science of nature. I therefore choose to be a person who is win-win in my way of thinking by accepting religion *and* science instead of being an either-or outcome thinker (i.e., win-lose). I personally believe the more science you learn and the more religion you incorporate, the more in awe you become of what our lives truly represent here on earth. It is all a gift to be used responsibly and enjoyed.

Interestingly, scientist Albert Einstein struggled with cultural teachings of religion as many have.[418] Although he was born a Jew, Einstein rejected the conservative theology of his prevailing culture and embraced a different concept of God that was more universal and less divided. It did not seem logical to him that a single God would look the way we think (although I could argue in a relative universe He would) and act the way presented in the mortal, translated Bible.[419] Yet Einstein, a man mostly of science, still believed in a subtler God figure that was more universal than we humans could comprehend and less exclusive than most current religious convictions. In fact, Einstein rarely tried to convince others that their religion was incorrect, as many of us so often do in today's society. His belief rather was one of a "God who reveals Himself in the harmony of all that exists."[420]

---

benefit the rest of our society (win-win outcomes).

[416] I do think religion is used as a tool this way in radical groups worldwide, especially in troubled youth.

[417] Einstein mused that coincidences are God's way of remaining anonymous.

[418] Einstein's birth country (Germany) persecuted him for his religious preferences, which later favored America's outcomes in World War II.

[419] I also personally struggle with the Old Testament in which God seems to be less forgiving and more wrathful, but humans need to have a certain balance of fear and safety in order to make appropriate moral choices (there needs to be an appropriate negative consequence of a bad choice for the feedback to work).

[420] Isaacson, *Einstein*, 551. Einstein was not really considered Christian either. I would point out the fact that we create God in our image (as a Him), or according to the Bible, it's the opposite—we are created in His image. I prefer to think we are all relative to his/her/its image depending on our own frame of reference, although it is easier to picture Michelangelo's Sistine Chapel version of God based

Unlike many religious followers who were zealots, Einstein never felt the urge to denigrate those who believed differently in God; he was tolerant of differing religions, and instead he tended to criticize *nonbelievers*. "What separates me from most so-called atheists is a feeling of utter humility toward the unattainable secrets of the harmony of the cosmos,"[421] he explained. In fact, Einstein tended to be more critical of debunkers, who seemed to lack humility or a sense of awe, than of the faithful. "The fanatical atheists," he wrote in a letter, "are like slaves who are still feeling the weight of their chains which they have thrown off after hard struggle. They are creatures who—in their grudge against traditional religion as the 'opium of the masses'—cannot hear the music of the spheres."[422]

While I prefer my own choice in religion, I cannot fault others in their choice, but I do believe some form of spiritual connection is a vital part of our life plan. I also believe it helps ground us into making the right choices in our lives. I believe in a mostly win-win world, where there are multiple rooms for multiple religions (and science) and where one does not negate or persecute[423] the findings or beliefs of another. And I do find it truly amazing that the complexities of the universe unfold before us every day in ways that in my mind can only mean order and godliness, however we choose to define it. Nothing happens by chance; it all happens for a reason. To really *get it*, you just need to look for it and truly see it, and submit to it. You and I may not always see the reason or hear the rhyme,[424] but there is a master plan, and we are all part of it.

---

on patterning. To a Chinese person or another ethnic frame of reference, you could see how these images of a Caucasian God and Jesus Christ would be counterintuitive to internal programming and would limit their willingness to accept this faith. Muslim faith on the other hand, in a way I think is more efficient for its people, does not promote the idea of idolatry or even displaying an image of God. The Muslim religion has efficiently doubled its growth in the United States over the last two decades, although this may simply reflect immigration and greater religious tolerances in our nation than in others.

[421] Ibid., 391.

[422] Ibid., 390.

[423] I say this because historically we have done this quite often.

[424] This is a subtle reference to what Jesus says in several biblical parables (he used these instead of a rhyme as his tool). "Do you have eyes but fail to see, and ears but fail to hear?" Mark 8:18 (NIV). "This is why I speak to them in parables: 'Though seeing, they do not see; though hearing, they do not hear or understand.'" Matthew 3:13 (NIV)

As eloquently summarized, "The human mind, no matter how highly trained, cannot grasp the universe. We are in the position of a little child, entering a huge library whose walls are covered to the ceiling with books in many different tongues. The child knows that someone must have written those books. It does not know who or how. It does not understand the languages in which they are written. The child notes a definite plan in the arrangement of the books, a mysterious order, which it does not comprehend, but only dimly suspects. That, it seems to me, is the attitude of the human mind, even the greatest and most cultured, toward God. We see a universe marvelously arranged, obeying certain laws, but we understand the laws only dimly."[425]

I am neither a great scientist nor a doctor of theology, and while I do not necessarily believe God is always directly involved in our common day-to-day personal lives, I do hope that he is for our sakes, and I cannot help but see evidence of his actions (directly or indirectly) everywhere around us when I train my eye to look for it. He has a plan, and it is mostly according to His will through Jesus Christ and the Holy Spirit, with our free will integrated in a way that hopefully pleases Him and honors Him. And while our anthropomorphic concept of God relative to us and how we see the universe may not be perfect,[426] I believe He has endowed us with the knowledge to learn from these few power-law principles that are created to help shape and regulate His universe. I believe that *these laws are there in order to guide us*, through our free choice, to do His will.

This is the way I therefore plan to live my own life using Christ as my guide and God as my absolute frame of reference. If you learn to rank Him highest in the pyramid of life, above yourself and even your family and country, I sincerely believe you will experience as I have that everything else blessedly falls into place like a well-designed game plan following a fortuitous roll of the dice. God provides all that we need in this game of life, if we learn to rank Him highest.

---

[425] Ibid., 386.

[426] The image of God (and of Jesus) in my mind was once again disproportionately influenced by an Italian master (among others)-Michelangelo.

PART 4

The Principle of Getting More for Less

"How shall I get it?
Given, Borrowed, Bought, Stolen?"

Getting What's Popular

My children have taught me more about life than I ever thought possible. In fact, once you have children, it seems hard to imagine how you could have ever enjoyed a life without them. They profoundly reshape your egocentric frame of reference, from one of me, myself, and I to a model that emphasizes family and community and focuses on others.

My children disproportionately account for so much of what I have come to value within my own life that it is only fitting that they be the subjects of the final few chapters, as they were my inspiration for most of this book. Their present choices and future actions follow similar principles of imbalance that I have cited throughout this work. It has been my goal to show them the nature of these relationships, as I feel these ideas are important and can be quantified and predicted.

If you asked each of my four children the nursery rhyme question, "How shall I get it?" (*it* could mean anything), you would see a naturally skewed imbalance in each of their answers. Similarly I want different outcomes for them based on their personalities, although I have a few common core desires for all of them. I also realize that many of their choices are inherently egocentric based on the physiological development of their immature brains. The selfless idea of what is best for others sometimes does not factor in until later in life.

One of my children would answer this question by choosing to not get much at all; rather she would selflessly *give* away everything she has in order to help others. I believe she was simply programmed this way from the start, and yes, we have continued to nurture this Christian behavior in her. A second daughter would buy everything physically in sight to feed her ego-driven desire for materialism and beauty if she were able. To her (for now) buying is the only way to acquire what she so selfishly desires (and yes, we have tried to help her unlearn this behavior), although she occasionally prefers to more efficiently *borrow and barter.* My third daughter seems to be content with a healthy balance of borrowing and buying and gifting and receiving. Finally, my son resorted to a more unproductive decision to *steal* to get what he wanted, which is something I would never have imagined from any of my four children. For all of them though, the choices still relate to the idea of *getting more for less.*[427] Once more life is imbalanced in ways that follow a simple rhyme, in this case without an obvious reason, but I am certain the rules still apply (in his case in a negative way unfortunately[428]).

My children's uniquely contrasting behaviors is what in fact inspired me to uncover these hidden principles. For example, the simple question, "How shall I get it?" is at the heart of the egocentric mind-set not only of young children but also to some degree our greater culture. Everyone tries to balance their self-interests with those of the greater culture, and it takes the power of time for us to change our frame of reference from one of egocentric thinking to one that focuses on the selfless interest of others. In a society that values and mostly teaches competition and self-interest, these outcomes unfortunately become mostly win-lose. In communities that favor cooperation and the larger population's interests, the outcomes are more often win-win. Learning this has been difficult in the context of American culture, especially since I have demonstrated to you how unequally we reward our population in many outcome measures.

---

[427] My son has a disease called addiction, and while I cannot relate to it on a personal level, it is similar to what we see medically, although this is hard for many to admit. His addiction is no different biochemically than alcoholism or other mental disorders, but it is really hard to accept as a father and a physician. Nonetheless, addiction is the ultimate idea of getting more for less, and you will do anything to feed that desire, including stealing from your own family.

[428] In his case, he wanted more money for less effort, so he made the decision to steal checks from us (as an adult) and forge our signature at the bank for several thousand dollars, which was unfortunately a felony charge. This, among a few other related bad choices, led to an unproductive outcome for him of jail time.

According to Zipf's model, how an individual chooses to focus on what he or she wants or needs relative to others is done to balance the efficiency of economies of multiple efforts and outcomes with the needs of the individual or the vital few. Similarly, learning that more cooperative behavior can collectively lead to more-efficient outcomes for the society (a win-win solution) ultimately benefits the individual as well, although this outcome is not always readily apparent to the person making the choices. While I have lived these principles throughout most of my life, and we have done so as a family, enumerating and better understanding these ideas has empowered me in ways I am only now beginning to fully comprehend.

I want my children to analytically think about their own choices, including *what is wrong and what is right* in relative terms to the ego ("I") versus another outside frame of reference (like God, family, or country). Stated another way that does not imply moral judgment, I want them to learn *a healthy balance* within this naturally imbalanced game of life. While I am no expert in game theory or in life, I do wish to show my children how our lives are cyclical human experiences of trying to learn right (e.g., win-win) versus wrong (e.g., lose-lose or win-lose) and make the right choices. There is no absolute right or wrong in a relativistic human world, which is something they need to learn for themselves, so in this context they will mostly see win-lose outcomes. This is something I hope to help them unlearn. But in a carefully chosen (or shaped) community that emphasizes Godly love and acceptance, the choices we make can lead to win-win situations. This is something I hope to model to them and get them to learn.

However we choose to define these concepts within the context of a fractal[429] frame of reference—as a nation of smaller communities in the case of America or as individuals in the case of my own family—the ideas

[429] Fractals are self-similar models that show the fractal nature of biology (e.g., branching of trees and rivers), anatomy (e.g., bronchial and arterial branching), and physiology (e.g., breathing and walking). Bruce J. West, "Fractal Physiology and the Fractional Calculus: A Perspective," *Frontiers in Physiology* 1, no. 12 (2010). In essence a fractal is a derivation of these power laws (when Benoit Mandelbrot first developed them, it was $z = z^2 + c$) and relates to similar models of human design efficiency, although the more common application is in fractal art seen on computers. The universe is made up of many self-similar reference frames that are fractal, each beautifully and efficiently designed by a master craftsman that appreciates these ideas of beauty and relativity.

are similarly related to the repeating but predictable skewed outcomes and this principle of learned efficiency (*more for less*).

Philosophically, I could argue that there is no absolute right or wrong choice for anyone. Everything is relative to someone else, somewhere else, and sometime else. But I also think I know in my heart what is right and what is wrong. It is internally programmed, and the two feel very different to me. Perhaps God is our only absolute reference, and learning what is right by understanding and getting closer to God is the challenge that defines all our lives. Perhaps this is the ultimate lesson we need to learn from our presence here and now.

Pareto would philosophically argue that what defines "right" is what is seen to represent the more common cultural norm. I would argue that in a morally efficient country, this would be true, and the greater good for more of the people would be a common goal for all of us. However, cultures vary so much based on their unique experiences that it is not my place to judge *what or who* is right or wrong. Furthermore, if I were truly honest, I would admit that I don't think we are morally efficient as a nation (and even worse as a planet), and I am afraid it is only going to get worse before it gets better again. It often tragically takes a catastrophe or a crisis (e.g., a war) for us to realize the gravity of matters and change our ways (i.e., our inputs or choices).

While I do not pretend to have a moral compass pointing due north, I do sincerely believe we can make choices that are *more* productive and repeatedly valued by society *and* morally acceptable to us as individuals (and that internally feel right). Learning this as an individual is difficult without some external frame of reference that is morally sound and socially acceptable (that's why we get educated by others, go to church and pray regularly, and try to have fair representation in government, and so on).

Similarly other choices seem discordant with our beliefs and are clearly less productive (and feel wrong collectively or as individuals). These choices may seem logical at times to the ego or the concept of what "I" want as an individual (for example, drug abuse), but they are often not conducive to healthy or happy lives in the long term if they are not compatible with the culture in which you live. Learning this through some acceptable frame of reference is part of the process of the human experience and exercising our free will (like moving away from geographies that are not supportive to our growth and positive development) and constitutes a large part of what drives the efficiency of the human machine.[430]

I believe that the measure of our human efficiency is no different than any other machine, whether artificial and man-made or godly. Human behavior seems like other measured universal efficiency for a reason.[431] Perhaps a 20–80 relation is a common theme among humans for a related reason as well, and this may help explain the repeated mechanical nature of the numerical frequency of how efficiently we learn these ideas and why at times what we value (e.g., income equality) seems *less* common.[432]

What's *common* in society is often likable by more people, but as I have demonstrated previously, it is not always best for us as a nation. People prefer common. You are more likely to interact with people who have common outcomes with you, to have common friends with those people, and to know others who have achieved similar goals and outcomes. So if common is cool within your frame of reference and you like being liked by many people who think you are cool and value the same common outcomes, then you will be popular. But often, it makes sense to think critically for

[430] Henderson's (power) law describes a concept known as experience curves, which are similar ideas (and differ from learning curves) and look graphically exactly the same as the other imbalance curves I have shown you. They relate similar ideas of efficiency in learning through experience.

[431] A Carnot engine, for example, classically performs at about 20 percent efficiency. The concept of a Carnot cycle involves a similar relationship between pressure and volume and thermodynamic work and efficiency, and these ideas led to the earthly concepts of pressure volume measurements in engines like the human heart pump, as well as the universal concepts of entropy, which relates to another concept called the arrow of time. They all derive from similar concepts of equilibrium mechanical principles, similar to the principle of imbalance Pareto described (also an equilibrium concept).

[432] My first idea for this book was to point out the repeating nature of this number 20, specifically its relative frequency in everything from income to marriages. But I think the more obvious explanation is that these power laws are all the same for everything involving human nature, including, as Zipf pointed out, our ability to communicate efficiently with words and numbers. Numbers often follow similar rules of distribution in society, a concept known as Benford's law (also a power-law distribution). In a universe that is all relative, the units of measure, including time and space, don't matter—each just measures some separation of one event from another either temporally or geographically, but in theory neither really exists in absolute terms.

yourself, to not follow common dogma, and to be unique, regardless of how cool or unpopular it may seem to the common culture.

One of my teenage girls currently thinks the more common way. She wants to seem cool and attractive to her peers, and it is very important for her to be liked by many people, although she knows she is loved by a vital few (her family and God). She is embarrassed, for example, about her adoption (she will not share that information with her peers) and her unique beauty, as she appears very different than the common teenage girl around her. In her case, she may be trying to fill a void created by her abandonment at birth by her biological parents, but whatever the case, the need to feel liked is constantly there, and as for many of us, it shapes her choices and actions.

My son also thinks in a similar way. He enjoys being cool with his common-thinking, uneducated peers to the point of making choices that interfere with his productive outcomes (for example, smoking marijuana, using other illegal drugs, and stealing). Even I admittedly enjoy the occasional thought of feeling a part of something cool: the city we lived in in West Virginia recently boasted the distinction of the "coolest small town in America." But often the context of what is common or cool needs to be considered in a larger frame of reference than our own.

Common as an outcome is not consistently valued by a society that ranks everything of value. Popular outcomes (as seen by population measures) in this case are seldom cool. Making the kind of money most people earn (i.e., the median income) in America, especially in West Virginia, is not cool. Living in the most uneducated state in America is also not cool. Abusing controlled substances for self-indulgence and pleasure is not cool, although increasingly common in our state and within America. Choosing to behave by doing what it is that "I" desire may not be the most productive outcome if it is not God's will and does not improve society.

All these outcomes consume your life disproportionately according to power-law relations and can negatively impact the lives of many others, including the individual, the family unit, and our nation as a whole. Living in a country with high divorce rates and high obesity does not lead to general societal well-being, despite these becoming the cultural norm. Having many superficial friends who hardly know you and rarely come to your aid in times of bad outcomes is rarely a recipe for meaningful relationships or social happiness. Living in a country that experiences moral and religious decay is rarely an effective choice that leads to happiness. I could similarly suggest here that win-lose outcomes are likewise not efficient and are therefore not productive toward societal happiness. Many of these outcomes I have described for the human condition involve win-lose outcomes, and for us to grow in the way we need to, win-win outcomes must become the norm. We must learn to think as Juran referred to with a "quality-mindedness."[433]

It is not all bad news, however, for our American children. There are some *common* core values that I believe are worth learning early in life that can be learned through proper choice and consequence (skewed) relationships—finding a few good friends that bring you up rather than down, developing a religious and spiritual belief that something more powerful than us helps guide our lives, discovering a career that magnifies your passions and benefits from your strengths, choosing a marriage or relationship that values effort and efficient communication, pursuing a valuable education with an emphasis on achievement and quality outcomes, and attracting yourself to a geography that embraces values similar to your own. All of these are similar concepts that rely on the understanding of the principles of human nature and involve establishing a sound and morally straight compass through your own frame of reference.

Using *More for Less* as a Principle to Reshape Our Personal Lives

On a basic level, my kids have seemingly always wanted more of what they like, which is one of the main reasons I chose the nursery rhyme I did from the beginning. On the surface, this rhyme emphasizes ideas with seemingly materialistic references, but within the substance of the reason for the rhyme is a message fundamentally much more valuable.

Like rhyming Daisy, my children want the same outcomes we all desire (such as simple happiness), and they want it as soon as possible and for

[433] Butman, *Juran*, 189.

little or no effort. My kids in fact embody the very idea of *getting more for less*, and what I have learned from the work of a few others (like Pareto, Zipf, and Juran) is that this idea is not so unreasonable if you understand the mechanics of human population outcomes.

Knowing the principles I've outlined in the previous chapters, it is not only *normal* but also even natural and potentially efficient to want to get more for less if you can think critically and understand the nature of the input-output relationship.

In theory *more* implies quantity (and less effort), but as shown in population outcomes, *more* (in quantity) is not always synonymous with better outcomes. There is a second component to the more-for-less principle: *quality and value*.

More for less as a principle (specifically emphasizing quality values with the 80–20 relation) is therefore a fundamental descriptor of human efficiency. It is programmed into the patterns we use to communicate as Zipf showed, is factored into the socioeconomic models that weirdly shape income and wealth as Pareto described, and disproportionately affects our lives through what we value as quality measures, as Juran demonstrated. This principle of getting more for less[434] weaves its way into everything we do that relates to human behavior, including basic human choice and consequence. Understanding this principle can help you in ways you never imagined.

We all in fact want the same thing—to be efficient in the choices we make so that we don't have to relearn and relive the same mistakes over and over that seem to characterize human history. How able we are to achieve these desired results are at the root of these issues—referred to as human ability in Pareto's work, human effort for Zipf, or human values for Juran. They all say the same thing to me, which is that we fundamentally all want what we want and that how we get there is dependent on many factors that relate to these repeating timeless principles known as power laws (including our thinking).

No matter what the reason for the outcomes, the principles always remain the same. Whether for income measures, geographic choices, career

---

[434] In my household I refer to this as GPL^3 (George's power law).

choices, educational attainment, job achievement, marriage, how we raise our children, how we incorporate religion into our family, our various and sundry contributions to society, or what we choose to rank and value within our own lives—the same principles of imbalance always hold true. A few of the choices we make (a vital few of the critical inputs) lead to imbalanced outcomes or consequences that can be shaped by understanding these principles. This is in part what I wish to teach my children about the idea of efficiency. If my children can learn which inputs count the most and achieve in these areas where choice is a factor, they can reshape the outcomes in ways that are more favorable to what they desire (a winning outcome solution).

Making the necessary changes in behavior to lead to efficient outcomes is much easier (takes less effort) for an individual than for an entire culture. This is as simple as making the individual choices from the nursery rhyme. While it may be somewhat more difficult based on your family of origin, your ethnic background, or your socioeconomic means and opportunity, it is nevertheless possible by making the appropriate choices. Those choices can involve something obvious like your education or something subtler like your choice in geography, marriage, or your basic values, like religion. Succeeding takes an understanding of these rules, some determination and resolve, some effort, and a fundamental internal value that you feel you must achieve.

If you value education, make it a priority, and do well by making more efficient choices, as education will disproportionately shape your present and future outcomes (including not only your wealth but also your health and how long you live). If you prefer athletics and are capable, make this your focus, and pour your heart into it, but try to also achieve some minimal level of education that is in keeping with the majority of the population and future dynamic needs. If you are interested in values that are consistent with timeless religious and moral principles, then attract yourself to them more often. If you aspire to be a career musician, do so while maintaining values that are healthy and in keeping with a productive society that values your input. Whatever you attach value to, attract more of it to you, by simply focusing on the vital few factors that are necessary to get you there.

Reshaping What Is Popular Culturally in America

Pareto, Zipf, and Juran all believed that human behavior (American or otherwise) is a repeating cyclical process, emphasizing the following general concepts: (1) behavioral choices and outcomes involve a predictable imbalance, (2) our results require some minimum effort (preferably through leaned efficiency), and (3) the skewed relationship begs for quality standards while emphasizing value within society. All three men's principles furthermore obey natural laws of the universe known as power laws.

Juran, the last survivor among the three (he died in 2008 at the age of 103), believed that most if not all the chronic problems that we face in society today—from environmental pollution to governmental waste and inefficiency to declining standards of education to an overcomplicated, inequitable health-care system—could be solved through principles of quality management.[435] I personally concur, and by using these few rules as power tools as he did, I believe we could achieve the desired win-win outcomes we all want within modern American culture.

The message from these core principles and from the men that espoused them was and still remains universal—from nineteenth-century Italy to twentieth-century postwar, industrial Japan to modern-day America. To quote Juran from one of his first formal lectures on his principles of quality controls, "Naturally, where we are dealing with the basic laws of nature, there are no differences between countries. The laws of gravitation and of thermodynamics apply with equal force [in Japan and in the United States]."[436]

Using a more relevant example from modern American culture, if we were interested in reshaping something as commonly influential as *income disparity*, we would see that most of the imbalance (70–80 percent) stems from a vital few causes, and by focusing on these few causal factors, we could make a large impact on restoring some balance to the naturally skewed system. While we will never restore a system of completely equal distribution of income, we could certainly shift the balance more favorably to benefit the

---

[435] Butman, *Juran*, 184–5. These would be win-win solutions for the individual and society.

[436] Ibid., 128. This was from a lecture in Japan to high-ranking business leaders.

masses by using some of these concepts. This is relevant because as I have suggested in previous chapters, income disparity seems linked to many other inequalities we value in American life, including general well-being.

Earlier, I asked of the top 20 percent (including myself),[437] "What do they know that we don't know? What is different about them, and why are they so special?" On the flip side of the same coin, "What is wrong with the other 80 percent and the more-common results that are so undesirable to our culture?"

The answers to these questions in part rest in awareness of these principles of imbalance (and their peculiar math), especially as they relate to what we currently value within America. Let me illustrate further with an example once more using *income* as the societal model.

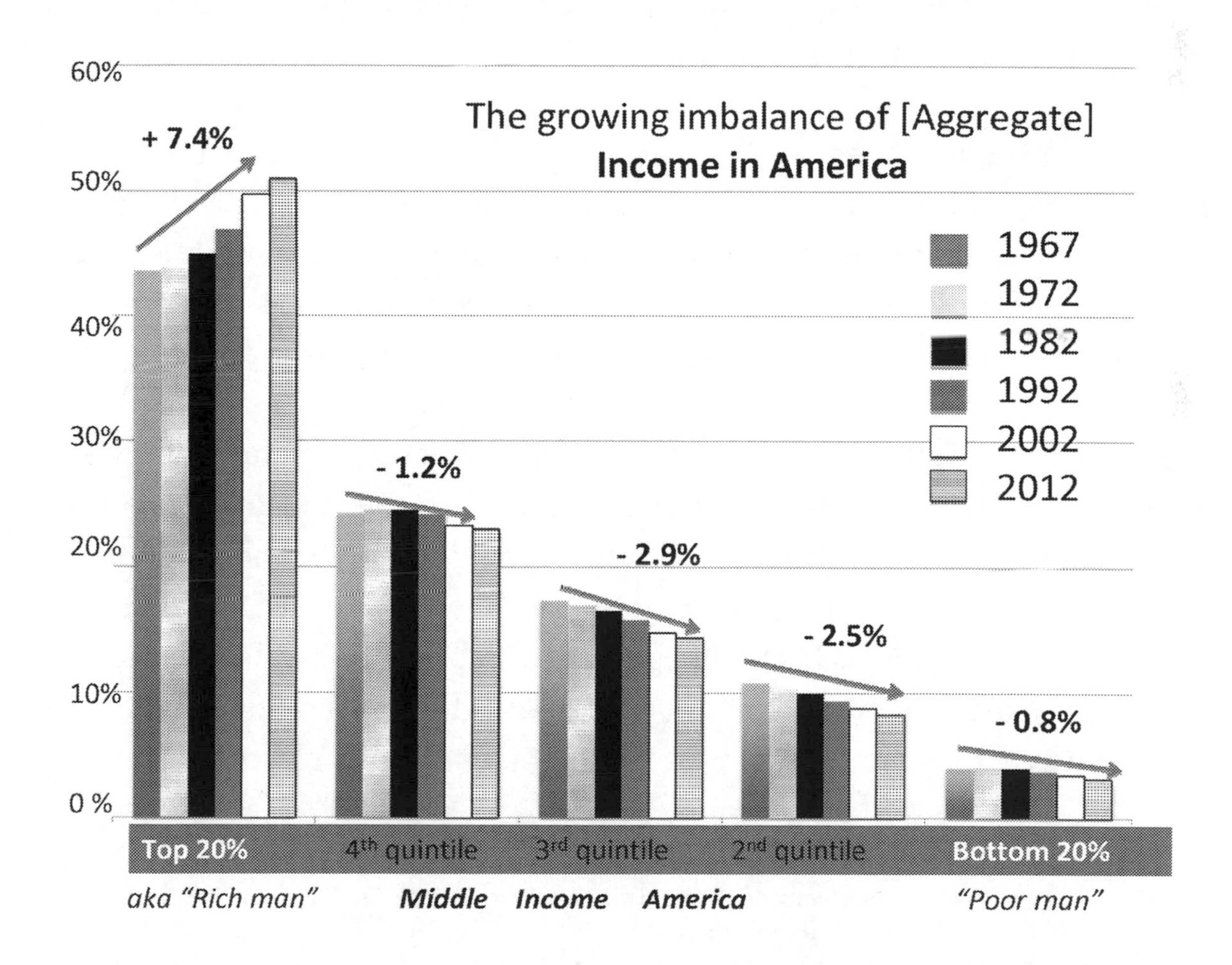

[437] I picked the top 20 percent due to the correlation to the 80-20 rule. But instead of targeting the top 20 percent of income earners, it makes sense to see what they have done differently on a statistical basis. Likewise, if you can figure out which 20 percent of the inputs account for 80 percent of the bad results, you can change those quality inputs.

Here I show the aggregate income of Americans over my lifetime.[438] This figure essentially shows which groups do well within America and which do not. As you can see, the top 20 percent do well consistently, and the other 80 percent do not do so well (from 1967 to now the top income group grew in income while every other group did not). We will never get rid of the top 20 percent, but we might be able to reshape this imbalance by analyzing the quintiles for the main causes of these imbalanced distributions. Forget the absolute amounts of income, and focus on which groups saw favorable change over the last forty years. Over my lifetime, most of the growth in aggregate income has only been in the top 20 percent of income earners, while most Americans have seen relatively negative growth or stagnation. So what do the top 20 percent know that most American's don't?

You may also observe from this image a secondary but important phenomenon known in game theory as zero-sum mathematics,[439] in which there are mostly winners and losers and the amount the winner keeps is at the loss of the losers. In the extreme case of zero-sum the winner takes everything, and the loser gets nothing. While this is not today's reality, the relative mathematical concept is similar. Think of income as an apple pie (which is an all-American pie): the bigger piece I get correspondingly makes your piece smaller. This is usually the case for resource allocation whether it is income or anything else (e.g., assets) of similar value. The amount the top 20 percent gained, while not huge, was offset by the other 80 percent of the entire population as losses of aggregate income. In the ideal world, we would more often achieve win-win outcomes rather than win-lose models, but a win-win outcome for something like income earnings in America (or

---

[438] US Census Bureau, Table H2: Share of Aggregate Income Received by Each Fifth and Top 5 Percent of Households, http://www.census.gov/hhes/www/income/data/historical/household/index.html.

[439] Not everyone would agree with me here, and I specifically would refer the interested reader to Robert Wright's great book *Non-Zero: The Logic of Human Destiny* (New York: Vintage, 2001). While things like improvements in technology and in medicine benefit everyone (win-win outcomes that arise thanks to a healthy balance between competitiveness and cooperation), I think he would still concur that money is allocated with zero-sum mathematics.

educational attainment) is much harder to control in the real world because of our human inefficiencies.[440]

How do we get more win-win situations rather than win-lose situations? It is not easy, and it involves learning to rethink the way we have been patterned to think about life. For most of us, life is a balance between competition and cooperation, and instinctually we struggle much like animals for what we believe to be limited resources (money, for example, in America or more simply food and water in poor countries[441]). This balance is what we sometimes call *a fine line*, though it's really a skewed curve. This balance is the polar opposite of the imbalances I've described throughout this book.

Instead of learning to become more competitive (and thus repeatedly experience win-lose outcomes as we have done cyclically throughout human history), we need to learn to be more cooperative (and get more win-win outcomes). Saying this is easy, but really understanding it takes a lot of effort, and not just one person but also an entire culture and ultimately the majority of the human race needs to understand it. For now, though, we are mostly continuing along on the win-lose path. With the power of time and awareness of these principles of imbalance, I hope most of us can change this.

The largest group (80 percent) of Americans has seen their share of aggregate income get smaller, while the top 20 percent have been consistently getting more, improving their relative quality of life. Why?

Yes, in part the imbalance is from how our country chooses to disproportionately value and reward certain professions beyond simple

[440] Even if we were able to achieve a win-win situation here in America, which is unlikely with income, the reality is someone else would be losing out somewhere else in the world. You cannot create income out of thin air for everyone (at least not for any lengthy period of time). Eventually the system restores itself to an equilibrium state that obeys these same laws of balance. So what is the solution? Accept that it is the way it is, and achieve what you are able to achieve, while recognizing it is a zero-sum game.

[441] My church (a few Methodists) completed a mission trip in Rwanda, for example, where water and food for HIV-positive orphans were luxuries. People in communities in Rwanda would beat and stone starving orphans for stealing to get what they needed to survive. In case you care to look, you will see the same power-law relationships of help and assistance provided to the poor and needy from a vital few nations and groups (mostly religious affiliations to help with evangelism). You will even see a skewed participation within the individual churches by what members contribute their time and efforts to the church activities. For example, I have not been a part of the vital few historically with my church but rather a part of the trivial many—but *I am still leaning*!

supply-and-demand relationships.[442] But in larger part the imbalance comes from a cultural shift in our moral and other social values, involving marriage and divorce, educational attainment, and religion. In part it comes from the fact that we place money ahead of almost every other value in American society, as evidenced by the correlation of income disparity with every measure of well-being in our culture.[443]

Through a simple shift of our values to more-efficient choices (and that foster cooperation rather than competition)—like better educational-attainment rates, more-successful and later marriages, successful cohabitation of multiple income earners, and equal opportunities[444] in careers and in education—we could efficiently expand the middle class in America rather than continue to watch it shrink as we have done over the last two generations. These ideas would not only be the more-efficient choices but also might even improve the moral nature of our country as well!

In my opinion, this *value shift* is one of the main reasons for the relative growth of household income in the top 20 percent. The top earners are not necessarily any smarter, do not necessarily work harder or more, and are not even luckier. The top earners have made some choices statistically on average that are more efficiently productive than other less desirable choices that are still valued in America. The top 20 percent of the country, for example, stay married more often (as a relative percentage) and for

---

[442] This I believe is a small part since these curves (personal income and occupational income) seem to be more exponential with less skew than household income. As economist Arthur Okun argued in his 1975 book, *Equality and Efficiency: The Big Tradeoff,* there is a reduced incentive to work and invest (and thus less growth) when incomes are more equally distributed. In the long-term perspective, however, others have demonstrated that more inequality is associated with less sustained growth. Andrew Berg and Jonathan Ostry, "Inequality and Unsustainable Growth: Two Sides of the Same Coin," *IMF Research Department Finance and Development* 48, no. 3 (September 2011).

[443] In *The Spirit Level*, Wilkinson and Pickett make a good argument that most everything we look at to measure well-being relates to the income disparity within our culture, suggesting to me that income has consumed too much of our efforts and our focus, similar to the alluded connotation from Daisy in the nursery rhyme. Income therefore is a good place to start if we want to efficiently improve our well-being. This is perhaps why President Obama has tried to emphasize this message as well.

[444] To Pareto, "equality of opportunity" refers to his belief that, in a healthy society, advancement must be opened to the superior members of all social classes (i.e., "meritocracy.") Charles Powers, *Vilfredo Pareto*, ed. Jonathan Turner, vol. 5 of *Masters of Social Theory* (Newbury Park, CA: Sage Publications, 1987), 22–3.

longer; they have achieved higher education on the whole (and in less popular professions that are usually in higher demand); the top quintiles live together in family units with multiple contributing members (either through efficient cohabitation or marriage).[445]

Most of America's population continues to make fewer of the "right" valued choices, while the top 20 percent have remained steadfast in their older and more-traditional values and better (more efficient) choices. It does not really matter to me how you perceive it; just understand the relative difference in the separation of these outcomes that America values.[446] The difference in values does not explain the natural imbalance for income—some imbalance always exists based on supply-and-demand principles—but it does help explain the growing separation in this imbalance that stems from our simple human behavior.

While the value shift does not explain the relative changes in weirdly skewed incomes of the super-elite (like the top 0.1 percent), I think this is mostly irrelevant, especially from an efficiency perspective. If I were among the governing elite, I would want to make policy changes that would (positively) affect 80 percent of the population and not 0.1 percent! Pareto would agree that to impact the most people, we would raise the mean income by mostly increasing the earnings of the disproportionately large number of low-income earners. The solution to achieving this outcome is to therefore *change the learned values* of the 80 percent, of the *under-performers*, and not

---

[445] For example, the top 20 percent (top quintile) has on average 2.0 income earners per household compared to the average of 1.3 (or stated another way, 78 percent of the top quintile have two or more income earners versus 38.8 percent for the nation). Similarly 78.2 percent of the top quintile households are married as opposed to only 48.7 percent of the average population (and compared to 16.7 percent of the bottom 20 percent being married couples). Carmen DeNavas-Walt, Bernadette D. Proctor, Jessica C. Smith, Table HINC-05: Income, Poverty, and Health Insurance Coverage in the United States: 2011–Current Population Reports, US Census Bureau, September 2012. In the top quintile 41 percent have bachelor's degrees or higher as opposed to 28 percent for the average in America. I could say these people are more efficient therefore in their marriages, and this is one of the main reasons they enjoy more income. The top 5 percent are even more efficient in these same measures of marriage and multiple income earners. This is a value we could choose to reward in our culture to promote it more often (that is an example of a win-win situation).

[446] If you were in the top 50 percent of occupations, the top 50 percent of achievers, the top 50 percent of marriages (i.e., not divorced), and in the top 50 percent of people who live within their means and save and so on, you would be in the top 1–5 percent by the power of the multiplicative math.

punish the more-efficient over-performers. In other words, redistributing income from the top 0.1 percent will not efficiently relieve the imbalance in my opinion, while changing the way the majority of Americans are patterned to think will be much more efficient and more effective and may indeed improve the quality of popular American life in the process.

In order to help the other 80 percent, we the top 20 percent have a few choices to make. The top 20 percent can choose to teach others what decisions are linked with which outcomes, thereby empowering others to choose to do for themselves—in part that is my intent with this book. We can also choose on our own to be more generous since we are blessed with better outcomes. We can gift back to society what we have been given. I think of it like tithing,[447] i.e., returning a tenth of what I earn to God (and in the case of giving more—alms). We could also as a society choose to disproportionately tax the *rich man* or redistribute wealth more actively by intervening with the process of income distribution, although I suspect this is the least efficient method.

In all likelihood, we will need to strike some balance between both concepts (robbing Peter to pay Paul) in order to satisfy the majority of the people within our skewed nation. This would be a win-win solution, although not very palatable to the elite. This is the cyclical nature of our history, and it also more simply becomes morally necessary when our own people are suffering to the degree this relative disparity has created. One fellow West Virginian that I am proud to quote here, Pearl S. Buck, who represents both our typical geography and the growing Daisys of our world, said, "If our American way of life fails the child, it fails us all."[448]

Many Americans probably want some sort of *distributive justice*[449] still existing within a system of capitalism, rather than an overhaul of

---

[447] One of my favorite verses from the Bible has to do with how we can as a nation return to God. "'Return to me, and I will return to you ... In tithes and offerings ... Test me in this,' says the LORD almighty, 'and see if I will not throw open the floodgates of heaven and pour out so much blessing that there will not be room enough to store it.'" From the Bible, Chapter of Malachi 3:7–10 (NIV)

[448] Pearl S. Buck, *Children for Adoption* (New York: Random House), 1965. She was also a Christian missionary much of her life.

[449] This idea (as believed by American economist John Roemer) helps remove the other associated ideas of socialism with redistribution of income and wealth. Roemer is by no coincidence a math major

the American system to one of true socialism. To do this, we will likely need to make some tax changes that preferentially disfavor the rich[450] and similarly change pay rates to disproportionately favor the poor (like increasing minimum wages). We will also need people who understand these concepts that can make these changes efficiently, and we will need to unlearn rewards that encourage minimal efforts and create a society of always expecting more for doing less.

Simply changing the way we compensate our highly educated will not work for any sustainable period, as they will not consider the return on their investment worth it in the long term, and attrition into these time-intense and degree-intense professions will follow natural supply-and-demand curves over time. Likewise, learning as a culture to make more-efficient choices in a world of growing globalization is critical if we wish to remain competitive (i.e., imbalanced in our favor).

No solution is easy, as both groups (top income earners and the remainder) are resistant to change. People have to want to change their own behavior for the outcomes to become more balanced (or less imbalanced) as a larger group, and no matter what you or I tell them or what our government tries to impose, the outcomes will not change significantly unless the majority of Americans want to change the current values in our society.

The reality is that it would not take much change in relative terms (maybe a few percent in the right direction), and I don't mean to imply that we should advocate a total redistribution of income for America or

---

from Harvard and espouses the same ideas of efficient choices in behavior that follows non-zero-sum game theory logic (i.e., cooperation and win-win outcomes).

[450] Zipf is one of the few who intelligently recognized that taxes are unfavorable to the poor unless they are progressive. In fact incomes would have to be decreased exponentially or logarithmically to have the same proportional effect. Likewise he showed how corporate taxes should be avoided since the net effect is a fixed tax rate, which is disproportionately felt by different corporations. Likewise in order to get a man to do more work, the increase in his compensation should be in proportion to a power of his income in order to achieve the desired effect (e.g., gratitude). Zipf, *Human Behavior and the Principle of Least Effort*, 467. Zipf refers to this as the exponential nature of incentives (it also explains the idea of social propinquity in which a person who can afford only five dollars for an evening's entertainment must seek a person in a socioeconomic class in which the five dollars will provide adequate incentive to go out). Similarly, most social interactions will tend to be within like socioeconomic classes, persons of similar means then live near one another, etc.

for the world. But a small change to only a few causes would restore some semblance of normalcy and get us back to where we have mostly been historically. I believe the only way we can achieve this is through principles that promote a change in our underlying values (e.g., education), as these changes will be magnified over a larger population *more* so than something that affects only the top 1–4 percent (*less*) of our population.

To further illustrate the idea of making more-efficient choices as a nation (and using the principle of getting more for less), when examining household income and its naturally skewed imbalance within American culture, I noticed the following variables disproportionately shaped income.[451]

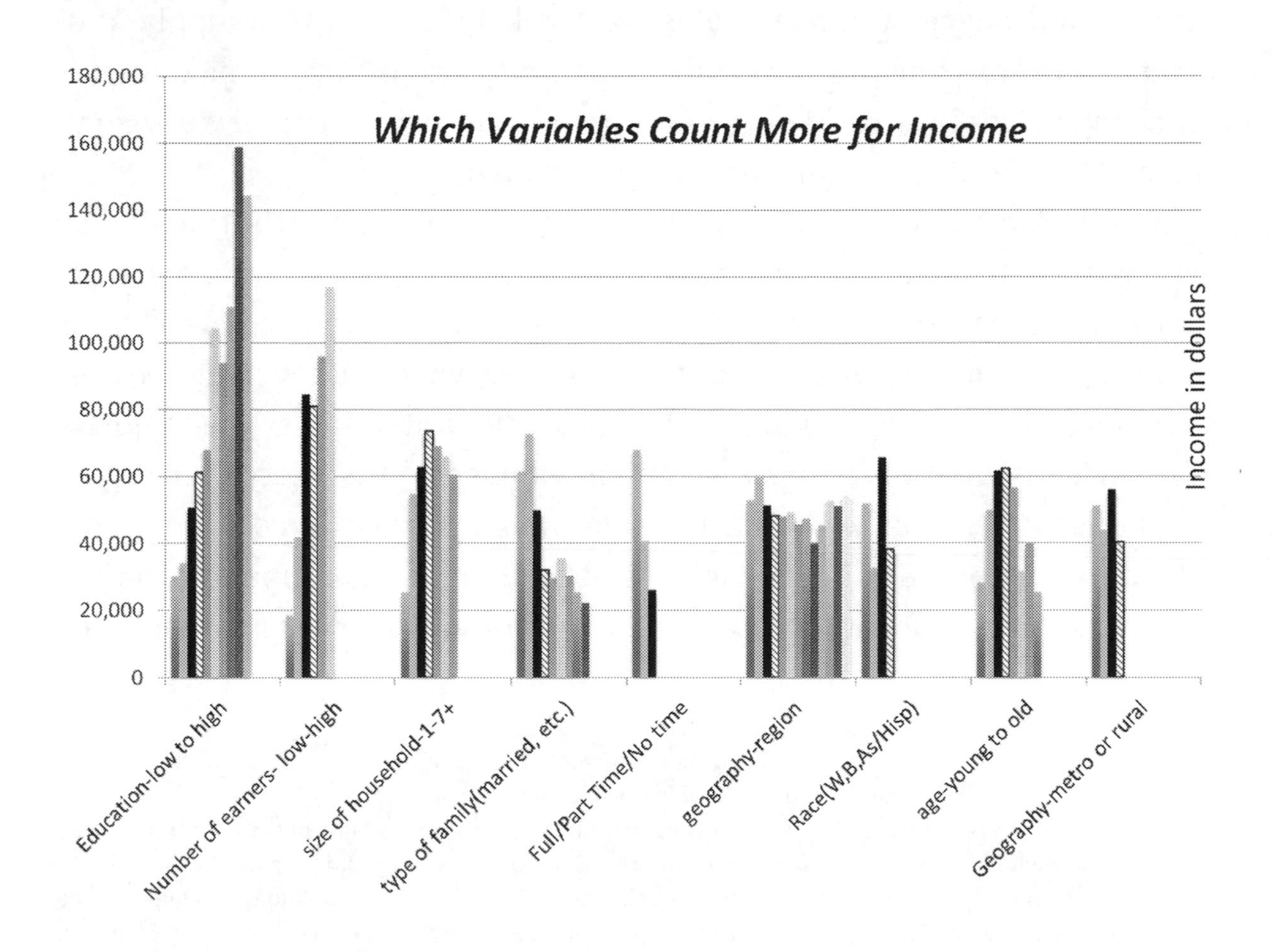

[451] This is all simply data from the US Census Bureau that show household income as a function of various inputs plotted in Excel and ranked in order from highest to lowest in effect. It is obvious to me where our efforts should be if this is a goal. I did not have data on religion versus none, but I would be curious to see if there is a correlation with our values and our success in America, as I would hypothesize.

The graph here shows how various factors affect income (median earnings of the American household unit, which is what I think of as the American family).[452] Notice here which factors account for most of the disparity in income for Americans: there are a few inputs that account for more disparity. Appropriate focus in these few areas would be efficient and productive use of our resources and efforts and would allow us to reshape the results more favorably over time.

The few inputs that seem most relevant are (1) educational attainment (more is better, even if it is just bachelor's degrees), (2) the number of people working in the home (the number of income earners), (3) the household size, and (4) the type of family (e.g., married). And in each case from above and throughout your life, you and society score yourself relative to others in these measures. Within each of these are additional factors such as gender and race (which can also be seen above), but since these are not easy variables to change, I do not choose to emphasize them here.[453] Your score is based on your choices and your outcomes, much like the roll of a die determines your collective result in a board game.

You'll also notice something very relevant here regarding the power of time. The outputs (values, educational attainment, etc.) that you achieve in the first portion of your life (i.e., your childhood) become the major factors that become the new inputs for the other portions of your life! All the outcomes that you eventually use to measure your success all relate back to these same childhood outcomes that became your adult inputs.

These areas where we score differently are areas where I believe we can make choices to improve the odds in our favor. For example, in my state, looking at household income, the average household number is lower than average (2.3 persons per household in our state versus 3.3 nationally), and

[452] US Census Bureau, Table HINC-01: Selected Characteristics of Households, by Total Money Income in 2011, http://www.census.gov/hhes/www/cpstables/032012/hhinc/hinc01_000.htm.

[453] What I infer from this is that it is hard to change what you are in these variables (you cannot change the fact you are Asian, for example, in the case of my daughter), although this is certainly a place where society can play a bigger role in equalizing opportunities to these otherwise immutable factors.

the levels of educational attainment are also lowest in the nation (fiftieth out of fifty). These would be a few areas where we could as a state more efficiently make a few changes to favorably increase our relative household income and thereby diminish disparity relative to other states. Likewise the type of family matters, as does working status (rather than choosing to remain unemployed or choosing to take advantage of disability benefits, as we consistently do as a state). This graphic also suggests that geography is not the main problem, at least not as directly, in terms of disparity for income regionally (the graph shows by region not by state). However, perhaps more indirectly, lower income in the case of our geography is a symptom of the other inputs, although still a factor in disparity of income.[454]

Reshaping American culture, while it sounds easy in principle, would clearly require some value changes, but this is part of my intended message, at least to my kids.[455] We could, for example, promote and reward educational attainment geographically through scholarship for work-study or through scholarships in exchange to stay and work locally. We could hold schools to higher accountability standards and link their allocations of money to efficiency principles and their ability to meet certain goals as outcomes. Likewise we could reward households with more income earners and intact marriages with some credit on taxes. We could have education tax credits as another example. But the idea is that we would be able to reshape the income disparity by making the right, and the more-efficient choices. Having said this, we can do only so much to educate people about the choices themselves, and we can do only so much to help them—the rest is up to them, their internal frames of reference, and their desire to individually change their own outcomes.

The most difficult challenge we face as a nation is our ability to change each other, but more relevantly helping each other to change *in ways that are favorable for all*. What we are able to change most efficiently is ourselves as individuals. Getting all or even most others to change is one of the

---

[454] A more proper analysis would include an analysis of the top 10 percent and bottom 10 percent for geography (or a Gini index), but the trends are what I am after here.

[455] It would be up to others with more power and influence than I have to make these changes in the hierarchal system of government and within our schools.

greatest challenges any society faces and is evident in every single one of the principles I have shown you in this book. Ask President Obama how hard it is to make change.[456] He made this idea the platform of his election campaign years ago, and he has battled these same cultural biases from the start, as people are resistant to external change more than change within themselves.[457]

As I have discovered with my own children, these ideas of changing them rely not only on them getting some minimal levels of education but also on developing quality core values (including religion), establishing trust with others, and unconditionally gifting love. To me, it means being more like Jesus and modeling that behavior to others. Understanding this is part of the challenge in life, but the rewards can be immense. In the parable of the sower, Jesus said, "Whoever has will be given more, and they will have an abundance. Whoever does not have, even what they have will be taken from them."[458] The only way, therefore, according to Him, is to enrich and develop your faith (through knowledge of the Father), and the rest I believe will naturally follow according to the principle of imbalance.

Having said that, some people (including my own children) must still learn through their own experiences and via their own values what does and does not work for them through feedback and proper analytical thinking.

Juran said it better than I, from his own observations in human relations, "In dealing with cultural patterns we are at our worst, hampered as we are by our limited basic knowledge and by our own emotional involvement."[459]

---

[456] Obama was a Harvard graduate as well. I would propose we need a leader with the mind-set of a tinker from Ivy League education and who understands the ideas of efficiency and game theory outcomes as well as strong values and sound morals, including religion and marriage. That person should logically be our next president.

[457] I don't mind coming to the conclusion that I need to do something differently, but I am inherently stubborn when someone thinks I should change in order to please him or her or because he or she knows better. My son is a great example of this, and people often need to come to these realizations on their own in order to find the impetus to change their behavior.

[458] Matthew 13:12 (NIV). This quote from the parable of the sower refers to not only hearing the word of God but also understanding it by having a good base or foundation from which to grow.

[459] Butman, *Juran*, 148. Pareto also referred to these same limitations shaped by emotions as rational and nonrational (emotional and sentimental), and they were related by the same imbalances we see with his income model. I could therefore suggest that income as an outcome is a complex measure of human nonrational sentimental behavior mostly balanced against a vital few key logical concepts,

A simple way to therefore shift the imbalance of income (for most households) in America is to do what the top 20 percent have been doing repeatedly over the last two generations: (1) get more education and stay educated (your education never really ends), (2) get married and preferably stay married, and (3) work, preferably with quality and core values in mind and using more than minimal effort. Keep doing more of the things that are valued by others, and together these factors become more than the simple sum of each one. Don't repeat the more-popular, inefficient outcomes within America, which are to remain uneducated, get divorced, and become unemployed! Don't do less and continue to selfishly expect more. You must match your work with some efficiency, and you must make the appropriate choices if you want more-desirable outcomes.

While it may sound odd to you that I preach efficiency with the fervor of a religion, I cannot help but credit these principles of efficiency and our related ability to reshape our outcomes based on where we are relative to other forms of life. Out of the hundreds of millions of life forms that have populated our planet through time, over 1 percent have survived through processes of efficiency in design and in choices. We are the fortunate ones (i.e. the 1 percent) for a reason, and it has a lot to do with our ability to reshape the world to efficiently fit our desires and needs.

As humans, we have become truly amazing creatures of efficiency. This efficiency is the main *reason* we exist and continue to survive within a universe that works by similar principles of efficiency, and learning this now (through the power of *a rhyme*) will help you disproportionately as you grow older and hopefully wiser.

Finally, one of the main influential factors in our human lives, and one you need to realize sooner rather than later, is your ultimate input—*time*. From the moment we are born (even to an unwed, poor teenage mother) and gifted the breath of life to the moment we leave our efficient DNA machinery behind, our life is a continuous time-decay process that becomes disproportionately faster the older we get. There is a time and season for

---

and learning to behave logically may be more efficient for you than behaving based on sentiment or emotions.

everything, just like there is a rhyme and reason for everyone. Learning to use your time efficiently—whether that is in your career, your marriage and family, your closeness to God, your dedication to helping others less fortunate than you, or simply enjoying leisure time—the concepts are all the same. Figure out how to get more of all these things early in life by defining and even reshaping your own frame of reference, and you will feel your earthly time is well spent. Figure this out in a way that glorifies your choice in God, and you will know eternal time and space with true bliss.

If you are asking yourself whether all this is doable and whether we have the necessary ingredients to make this happen, through our choices and through our free will, I would answer using a final quote from the current leader of the free-thinking world (and who is no doubt a power thinker and who gets the laws I have described here), President Obama, our current chief (and a lawyer):

"Yes we can."

In conclusion I believe these few principles work always and everywhere and are general life skills worth realizing. They are universal and describe outcomes for everything human and nonhuman that involves any process of efficiency. Learning these power relationships is therefore critical if you want to reshape life the way you want it to be. Learning and using the *principle of getting more for less* is the easiest way to get you there. Therefore, use your time wisely to find your own frame of reference in this life of never-ending relativity. Make your frame of reference one of value not only to you, to your family, and to other Americans, but more importantly to God, and you too can and will be richly rewarded and blissfully happy.

References

American Cancer Society. "US Cancer Incidence Rates by Site and State." *Cancer Facts and Figures 2011*. Atlanta: American Cancer Society, 2011.

Adamic, Lada A., and Bernardo A. Huberman. "Zipf's Law and the Internet." *Glottometrics* 3 (2002): 143–50.

Anderson, Chris. *The Long Tail: Why the Future of Business Is Selling Less of More*. New York: Hyperion Publishing, 2006.

Barabasi, Albert-Laszlo. *Linked: How Everything Is Connected to Everything Else and What It Means for Business, Science, and Everyday Life*. New York: Basic Books, 2014

Baum, Sandy, Jennifer Ma, and Kathleen Payea. *Education Pays 2010: The Benefits of Higher Education for Individuals and Society*. College Board Advocacy & Policy Center. http://trends.collegeboard.org.

Bentley, R. Alexander, Paul Ormerod, and Michael Batty. "An Evolutionary Model of Long Tailed Distributions in the Social Sciences." Cornell University Library, http://arxiv.org/abs/0903.2533v1 (2009)

Berg, Andrew, and Jonathan Ostry. "Inequality and Unsustainable Growth: Two Sides of the Same Coin." *IMF Research Department Finance and Development* 48, no. 3 (September 2011).

Bolender Initiatives. "Vilfredo Pareto." Accessed August 21, 2014. www.bolenderinitiatives.com/sociology/Vilfredo-pareto-1848-1923.

Bowen, William G., Matthew M. Chingos, and Michael S. McPherson. *Crossing the Finish Line: Completing College at America's Public Universities.* Princeton: Princeton University Press, 2011.

Buchanan, Mark. *Ubiquity: Why Catastrophes Happen.* New York: Broadway Books, 2002.

Butman, John. *Juran: A Lifetime of Influence.* New York: Wiley and Sons, 1997.

Caxton, William. *The Game and Playe of the Chesse.* Edited by Jenny Adams. Kalamazoo, MI: Medieval Institute Publications, 2009. http://d.lib.rochester.edu/teams/publication/adams-caxton-game-and-playe-of-the-chesse. (Caxton's book, which was originally published in 1474, is based on a book by Jacopo Da Cessole.)

Chiang, Ming-Chang, et al. "Hierarchal Clustering of the Genetic Connectivity Matrix Reveals the Network Topology of Gene Action on Brain Microstructure: An N=531 Twin Study." Accessed at http://users.loni.usc.edu/~thompson/PDF/ISBI2011/MC-GeneClust-ISBI11.pdf, 2011

Christiadi. *Population Projection for West Virginia Counties*. Bureau of Business and Economic Research and College of Business and Economics—West Virginia University, August 2011.

Clauset, Aaron, Cosma Rohilla Shalizi, and M. E. J. Newman. "Power-Law Distributions in Empirical Data." *SIAM Review* 51, no. 4 (2009): 661–703. http://www.santafe.edu/research/publications/sfi-bibliography/detail/?id=1826.

Copen, Casey E., et al. "First Marriages in the United States: Data from the 2006–2010 National Survey of Family Growth." *National Health Statistics Report* 49 (March 22, 2012).

DeNavas-Walt, Carmen, Bernadette D. Proctor, and Jessica C. Smith. *Income, Poverty and Health Insurance Coverage in the United States: 2011*. US Census Bureau, September 2012. http://www.census.gov/topics/income/publications.html

Diamond, Jared. *Guns, Germs, and Steel: The Fates of Human Societies*. New York: W.W. Norton and Company, 1997.

Dickens, Charles. *A Tale of Two Cities from Charles Dickens: Four Complete Novels*. New York: Weathervane, 1982.

Darwin, Charles. *The Origin of Species*. 1859. Reprint, New York: Random House, 1979.

Dawkins, Richard. *The Selfish Gene*. New York: Oxford University Press, 1976.

Edgeworth, Y. *Mathematical Psychics: An Essay on the Application of Mathematics to the Moral Sciences*. 1881. http://socserv.mcmaster.ca/~econ/ugcm/3ll3/edgeworth/mathpsychics.pdf.

Ferriss, Timothy. *The 4-Hour Body.* New York: Crown Archetype, 2010.

Friedman, Thomas L. *The World Is Flat: A Brief History of the 21st Century.* Solon, OH: Findaway World, 2005.

Gabaix, Xavier. "Power Laws in Economics and Finance." *Annual Review of Economics* 1 (2009): 255–93. http://www.annualreviews.org/journal/economics.

Gardner, Howard. *Frames of Mind: The Theory of Multiple Intelligences.* New York: Basic Books, 1983.

Gladwell, Malcolm. *The Tipping Point: How Little Things Can Make a Big Difference.* New York: Back Bay Books, 2002.

Gould, Stephen Jay. *Questioning the Millennium: A Rationalist's Guide to a Precisely Arbitrary Countdown.* New York: Harmony Books, 1997.

Hart, David M. "Life and Works of Gustave de Molinari." Library of Economics and Liberty. http://www.econlib.org/library/Molinari/MolinariBio.html.

Hoyert, Donna. "75 Years of Mortality in the United States." *NCHS Data Brief,* no. 88 (March 2012).

Hymowitz, Kay, Jason Carroll, W. Bradford Wilcox, and Kelleen Kaye. *Knot Yet: The Benefits and Costs of Delayed Marriage in America.* The Relate Institute, the National Marriage Project at the University of Virginia, and the National Campaign to Prevent Teen and Unplanned Pregnancy, 2013.

Isaacson, Walter. *Einstein: His Life and Universe.* New York: Simon and Schuster, 2007.

James, A., and M. J. Plank. "On Fitting Power Laws to Ecological Data." 2007. http://arxiv.org/pdf/0712.0613v1.pdf.

Julian, Tiffany. *Work-Life Earnings by Field of Degree and Occupation for People With a Bachelor's Degree: 2011 American Community Survey Briefs.* US Census Bureau, October 2012. http://www.census.gov.

Juran, Joseph M. *Bureaucracy, A Challenge to Better Management: A Constructive Analysis of Management Effectiveness in the Federal Government.* 1944.

———. *Quality Control Handbook.* New York: McGraw-Hill, 1951.

Kingkade, Tyler. "Student Loans: New Survey Finds College Grads Carry Large Debt into Middle Age." *Huffington Post,* May 2, 2012. http://

www.huffingtonpost.com/2012/05/02/student-loans-college-debt_n_1468831.html.

Kleiber, Max. "Body Size and Metabolism." *Hilgardia* 6, no. 11 (1932): 315–51.

Koch, Richard. *The 80/20 Principle: The Secret to Achieving More with Less.* New York: Crown Publishing, 2008.

Krauss, Lawrence M. *A Universe from Nothing: Why There is Something Rather Than Nothing.* New York: Free Press, 2012.

Levine, Arthur, and Diane R. Dean. *Generation on A Tightrope: A Portrait of Today's College Student.* San Francisco: Jossey-Bass, 2012.

Lindert, Peter, and Jeffrey Williamson. "American Incomes Before and After the Revolution. "http://gpih.ucdavis.edu/files/w17211.pdf (July 2011)

Lockard, C. Brett, and Michael Wolf. "Employment Outlook: 2010–2020: Occupational Employment Projections to 2020." *Monthly Labor Review* (January 2012): 84–108.

Lui, Julian C., and Jeffrey Baron. "Mechanisms Limiting Body Growth in Mammals." *Endocrine Reviews* 32, no. 3 (June 2011): 422–40.

Malthus, Thomas. *An Essay on the Principle of Population.* London: J. Johnson Publishers, 1798. Accessed at http://www.econlib.org/library/Malthus/malPop.html

Mandelbrot, Benoit. *The Fractal Geometry of Nature.* New York: W. H. Freeman and Company, 1982.

Marvin, Donald. "Occupational Propinquity as a Factor in Marriage." PhD diss., University of Pennsylvania, 1918.

Maslow, Abraham. "A Theory of Human Motivation." *Psychological Review* 50, no. 4 (1943): 370–96. http://psychclassics.yorku.ca/Maslow/motivation.htm.

National Center for Education Statistics. Table 313: Undergraduate Degree Fields. http://nces.ed.gov/programs/coe/indicator_cta.asp#info.

National Institute on Drug Abuse. *Preventing Drug Use among Children and Adolescents: A Research-Based Guide for Educators, Parents and Community Leaders, 2002.* 2nd ed. Bethesda: National Institutes of Health, 2003.

The National Marriage Project and Institute for American Values. *The State of Our Unions.* Charlottesville, VA: National Marriage Project and Institute for American Values, 2005.

The National Marriage Project and Institute for American Values. *The State of Our Unions: Marriage in America 2012*. Charlottesville, VA: National Marriage Project and Institute for American Values, 2012.

Newman, M. E. J. "Power Laws, Pareto Distributions and Zipf's law." *Contemporary Physics* 46, no. 5 (2005): 323–51.

Norton, Michael, and Dan Ariely. "Building a Better America—One Wealth Quintile at a Time." *Perspectives on Psychological Science* 6, no. 9 (2011): 11. pps.sagepub.com.

OECD. *Education Today 2013: The OECD Perspective.* OECD Publishing, 2012. http://dx.doi.org/10.1787/edu_today-2013-en.

———. *How's Life? Measuring Well-Being*. Paris: OECD Publishing, 2011. http://dx.doi.org/10.1787/9789264121164-en.

———. *OECD Fact Book 2011: Economic, Environmental and Social Statistics*. OECD Publishing, 2011. http://dx.doi.org/10.1787/factbook-2011-en.

———. *Understanding the Brain: The Birth of a Learning Science*. Paris: OECD Publishing, 2007. http://dx.doi.org/ 10.1787/9789264029132-en.

Okun, Arthur. *Equality and Efficiency: The Big Tradeoff*. Washington, DC: Brookings Institution Press, 1975.

Pareto, Vilfredo. *The Mind and Society*. Translated and edited by Arthur Livingston. San Diego: Harcourt, Brace and Company, 1935.

Pareto, Vilfredo. "New Theories of Economics." *The Journal of Political Economy* (1896).

Persons, Stow. *American Minds: A History of Ideas*. New York: Holt, Rinehart, and Winston, 1958.

Pew Research Center. "The Decline of Marriage and Rise of New Families." November 18, 2010. http://www.pewsocialtrends.org/2010/11/18/the-decline-of-marriage-and-rise-of-new-families/2/#ii-overview.

Powers, Charles. *Vilfredo Pareto*. Edited by Jonathan Turner. Vol. 5 of *Masters of Social Theory*. Newbury Park, CA: Sage Publications, 1987.

Ryan, Camille L., and Julie Siebens. *Educational Attainment in the United States: 2009*. US Census Bureau, February 2012.

Rand, Ayn. *Atlas Shrugged.* New York: Random House, 1957.

———. *The Fountainhead.* Indianapolis: Bobbs Merrill, 1943.

Rousseau, Jean-Jacques. *The Social Contract and Discourse on the Origin of Inequality*. Edited by Lester G. Crocker. New York: Washington Square Press, 1967.

Samaras, Thomas T. *Human Body Size and the Laws of Scaling: Physiological, Performance, Growth, Longevity and Ecological Ramifications*. Hauppauge, NY: Nova Science, 2007.

Sandel, Michael J. *What Money Can't Buy: The Moral Limits of Markets*. New York: Farrar, Straus and Giroux, 2013.

Siegfried, Tom. *A Beautiful Math: John Nash, Game Theory, and the Modern Quest for a Code of Nature*. Washington DC: Joseph Henry Press, 2006.

Silva, A. Christian, and Victor M. Yakovenko. "Temporal Evolution of the 'Thermal' and 'Super Thermal' Income Classes in the USA during 1983–2001." *Europhysics Letters* 69, no. 2 (2005): 304–10.

Simkin, M. V., and V. P. Roychowdhury. *Re-inventing Willis. Physics Reports* 502, no. 1 (2011): 1–35.

Smith, Adam. *An Inquiry into the Nature and Causes of the Wealth of Nations*. 1776. http://www2.hn.psu.edu/faculty/jmanis/adam-smith/Wealth-Nations.pdf.

Snider, Susannah. "Best Values in Public Colleges, 2013." *Kiplinger's Personal Finance*. February 2013. http://www.kiplinger.com/article/college/T014-C000-S002-best-values-in-public-colleges-2013.html#yl4cBcLGP2f1iOGX.99.

"The State of American Well-Being 2013: Gallup-Healthways Well-Being Index." Healthways and Gallup Consulting Firm. Accessed August 23, 2014. http://info.healthways.com/wbi2013.

Tambovtsev, Yuri, and Colin Martindale. "Phoneme Frequencies Follow a Yule Distribution (The Form of the Phonemic Distribution in World Languages)." *Journal of Theoretical Linguistics* 4, no. 2 (2007): 1–11.

Thornton, James. "Vilfredo Pareto: A Concise Overview of His Life, Works and Philosophy." http://jkalb.freeshell.org/misc/pareto.html.

US News and World Report. "Best Colleges 2013." http://colleges.usnews.rankingsandreviews.com/best-colleges.

US Census Bureau. Current Population Survey. *2011 Annual Social and Economic Supplement.* http://www.census.gov/hhes/www/cpstables/032011/hhinc/hinc01_000.htm.

———. "Education." *Statistical Abstract of the United States: 2012.*

———. *Fact Sheet: 2006–2008 American Community Survey 3-Year Estimates.*

———. *Statistical Abstract of the United States, Income, Expenditures, Poverty and Wealth.* 2012

———. Table 706: Household Income—Distribution by Income Level and State: 2009. *Statistical Abstract of the United States, 2012.*

Waldrop, Judith, and Sharon M. Stern. *Disability Status: 2000.* US Census Bureau, March 2003.

Wikipedia. s.v. "Thomas Robert Malthus." Last modified August 17, 2014. http://en.wikipedia.org/wiki/Thomas_Robert_Malthus.

Wikipedia. s.v. "Tinker, Tailor." Last modified August 7, 2014. http://en.wikipedia.org/wiki/Tinker,_Tailor,_Soldier,_Sailor.

Wilkinson, Richard, and Kate Pickett. *The Spirit Level: Why Greater Equality Makes Societies Stronger.* London: Allen Lane Publishing, 2009.

Wright, Robert. *Non-Zero: The Logic of Human Destiny.* New York: Pantheon Books, 2000.

Zipf, George Kingsley. *Human Behavior and the Principle of Least Effort: An Introduction to Human Ecology.* Eastford, CT: Martino Publishing, 2012. First published 1949 by Addison Wesley Press.

Appendix

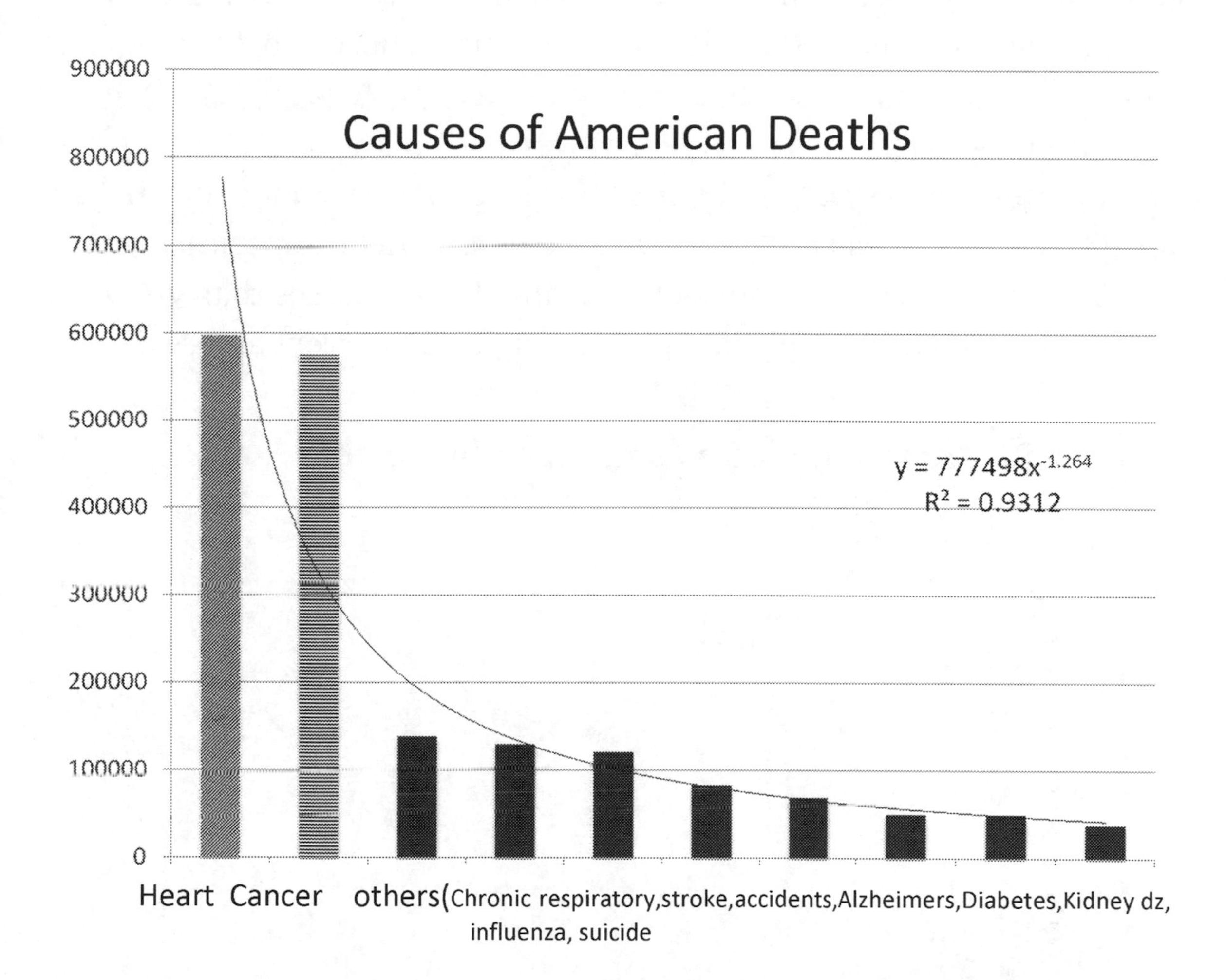

Image 1A: This graph shows the ten most-common causes of mortality (and morbidity) in America (from CDC data 2010). Certain diseases are responsible for the majority of deaths of people in our society and worldwide. This is no coincidence since humans choose the same inputs and have similar biological (efficient) machines. For the United States here and now, this graph shows the top ten causes of death versus absolute numbers of people.

There are a few vitally important diseases that kill most people (they are the efficient killers) and some less-common causes, and this has of course changed over time to reflect changes in our technology and improvements in public health and medicine. That is not to say of course that you couldn't die of some other rare cause, but the stats suggest that for most of us, one of these ten will be responsible. A few causes (20 percent) kill most people, so this is where more efforts would yield better improvements for more people statistically. National Vital Statistics reports from Centers for Disease Control and Prevention and accessed at http://www.cdc.gov/nchs/deaths.htm and publication by Donna Hoyert, "75 Years of Mortality in the United States," *NCHS Data Brief*, no. 88 (March 2012). Sadly suicide has evolved as a top-ten cause of death in the modern era for America. I would point out that we are more efficient at curing these illnesses than we were in the past, and our life expectancy is up roughly 30 percent as a result.

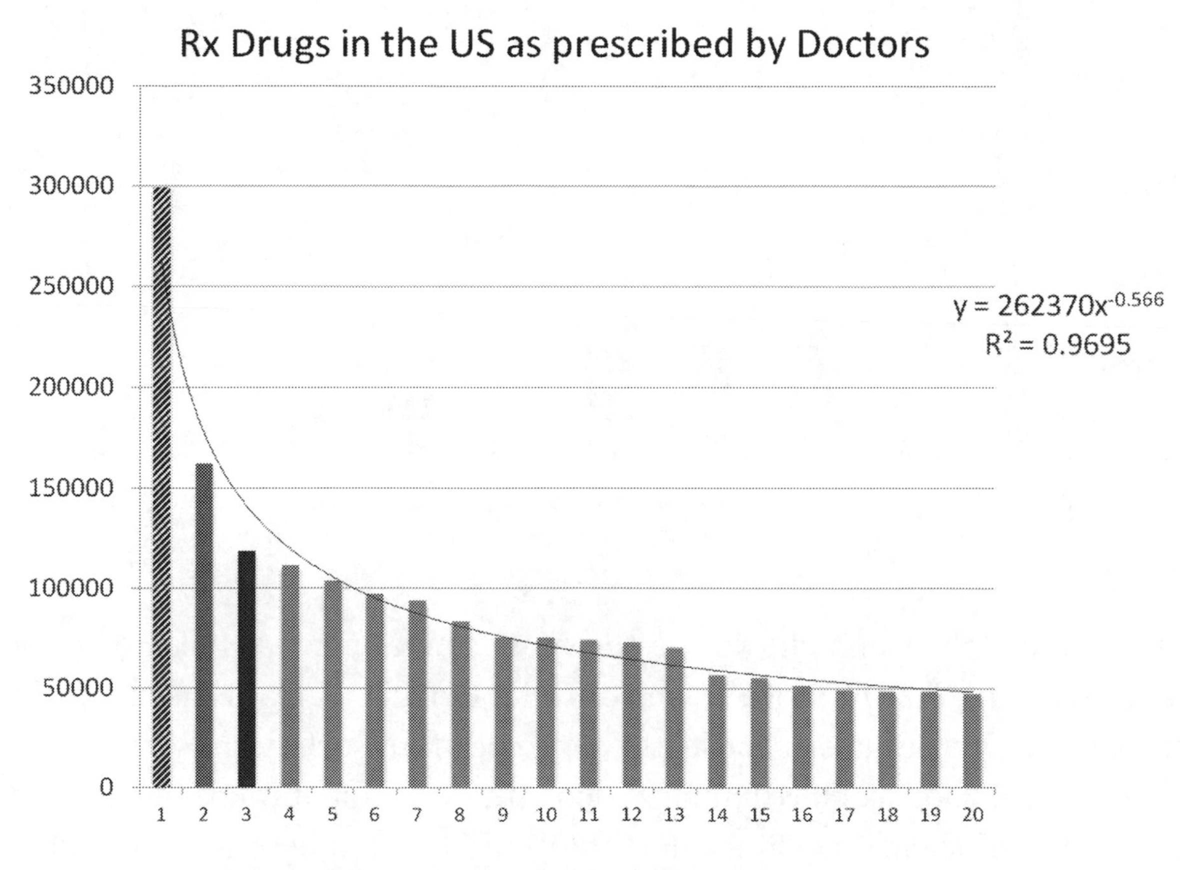

Image 2A: This image shows CDC data for the most-common prescriptions in the United States. This data fits a power law nicely. I would think this is because of the relation I showed with heart disease and cancer and other illnesses, since pain medications are common in cancer and anticholesterol medicine common in heart disease. However, somehow I think it is more than these illnesses and is a statement about our society. We prefer easy pills for pain and suffering (mood elevators). The top twenty drugs together account for 70.5 percent of *all* drugs used in the prescription medical world.

	College Majors that Pay 2012	*starting salary*	*mid- career salary*
1	Petroleum Engineering	$98,000	$163,000
2	Aerospace Engineering	$62,500	$118,000
3	Actuarial Mathematics	$56,100	$112,000
4	Chemical Engineering	$67,500	$111,000
5	Nuclear Engineering	$66,800	$107,000
6	Electrical Engineering (EE)	$63,400	$106,000
7	Computer Engineering (CE)	$62,700	$105,000
8	Applied Mathematics	$50,800	$102,000
9	Computer Science (CS)	$58,400	$100,000
10	Statistics	$49,300	$99,500
11	Physics	$51,200	$99,100
12	Mechanical Engineering (ME)	$60,100	$98,400
13	Biomedical Engineering (BME)	$54,900	$98,200
14	Government	$42,000	$95,600
15	Economics	$48,500	$94,900
16	International Relations	$40,600	$93,000
17	Materials Science & Engineering	$60,100	$91,900
18	Industrial Engineering (IE)	$59,900	$91,200
19	Software Engineering	$59,100	$90,700
20	Environmental Engineering	$47,900	$89,700

Table 1: College majors that pay in 2012 ranked in order of top twenty using midcareer earnings as the outcome measure (and looking at bachelor's-degree holders only). This is another reference from www.payscale.com from the last two years.

year	*1990*	*2001*	*2008*
Population(x1000)	175440	207983	228182
Christian	151225	159514	173402
Jewish	3137	2837	2680
Muslim	527	1104	1349
Buddhist	404	1082	1189
Unitarian	502	529	586
Hindu	227	766	582
Native American	47	103	186
Sikh	13	57	78
Wiccan	8	134	342
Pagan	NA	140	340
Spiritualist	NA	116	426
Other			
Non-classified	991	774	1030
No religion	14331	29481	34139
No answer	4031	11246	11815

Above as %	*1990*	*2001*	*2008*
Christian	0.861975604	0.766956915	0.759928478
Jewish	0.017880757	0.013640538	0.011745011
Muslim	0.003003876	0.005308126	0.005911947
Buddhist	0.002302782	0.005202348	0.005210753
Unitarian	0.002861377	0.002543477	0.002568125
Hindu	0.00129389	0.003682993	0.002550596
Native American	0.000267898	0.000495233	0.000815139
Sikh	7.40994E-05	0.000274061	0.000341832
Wiccan	4.55996E-05	0.000644283	0.001498804
No religion	0.081686047	0.141747162	0.149613028
Refuse to answer	0.022976516	0.070501649	0.068136469

RELIGIONS OF AMERICA FROM 1990 to 2008 from US Census Bureau Report

Table 2: Religions in America from 1990 to present. US Census, Table 75: Self-Described Religious Identification of Adult Population: 1990, 2001, and 2008, *Statistical Abstract of the United States: 2012*, 61. Numbers are in thousands, so 175,440 represents 175,440,000. The methodology of the American Religious Identification Survey (ARIS) 2008 replicated the methods used in previous surveys. The three surveys are based on random-digit-dialing telephone surveys of residential households in the continental

United States (forty-eight states): 54,461 interviews in 2008, 50,281 in 2001, and 113,723 in 1990. Respondents were asked to describe themselves in terms of religion with an open-ended question. Interviewers did not prompt or offer a suggested list of potential answers. Moreover, the self-description of respondents was not based on whether established religious bodies, institutions, churches, mosques, or synagogues considered them to be members. Instead, the surveys sought to determine whether the respondents regarded themselves as adherents of a religious community. Subjective rather than objective standards of religious identification were tapped by the surveys.

LIFE SATISFACTION by GDP per capita (a measure of Wealth)

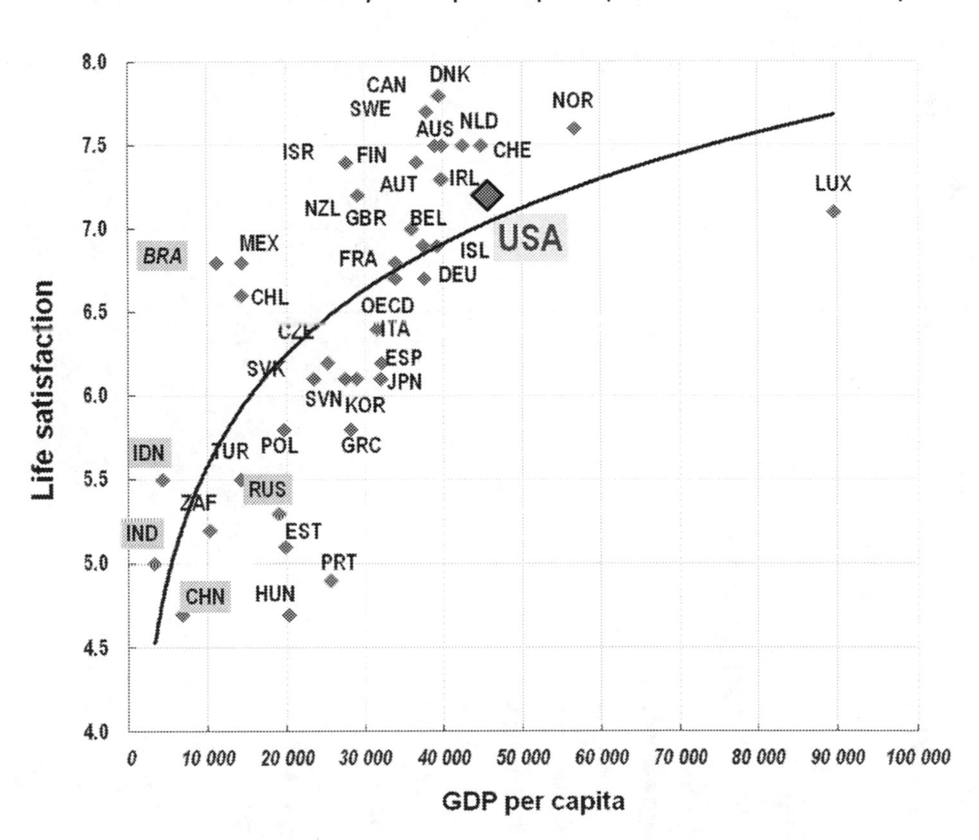

Cantril Ladder (from 2010 data)

Image 3A: This is known as the *Cantril Ladder* and plots country wealth (the input) versus life satisfaction (the output). It shows a few interesting things

from a lot of countries. Notice the relatively high rank of the United States toward the top right (shown as an enlarged diamond). As you move along the x-axis from left to right and as individual countries on average make more money per citizen (GDP per capita), the feeling of life satisfaction goes up sharply to a point and then plateaus. More is better up to a point! Then the result becomes less efficient. That is, the payout becomes more blunted, implying other factors are important. This was reprinted from the OECD with permission and does not imply any endorsement of my ideas in this book by their organization. Reference http://dx.doi.org/10.1787/888932493651 from OECD publication *How's Life: Measuring Well-Being* (Paris: OECD Publishing, 2011), 273.

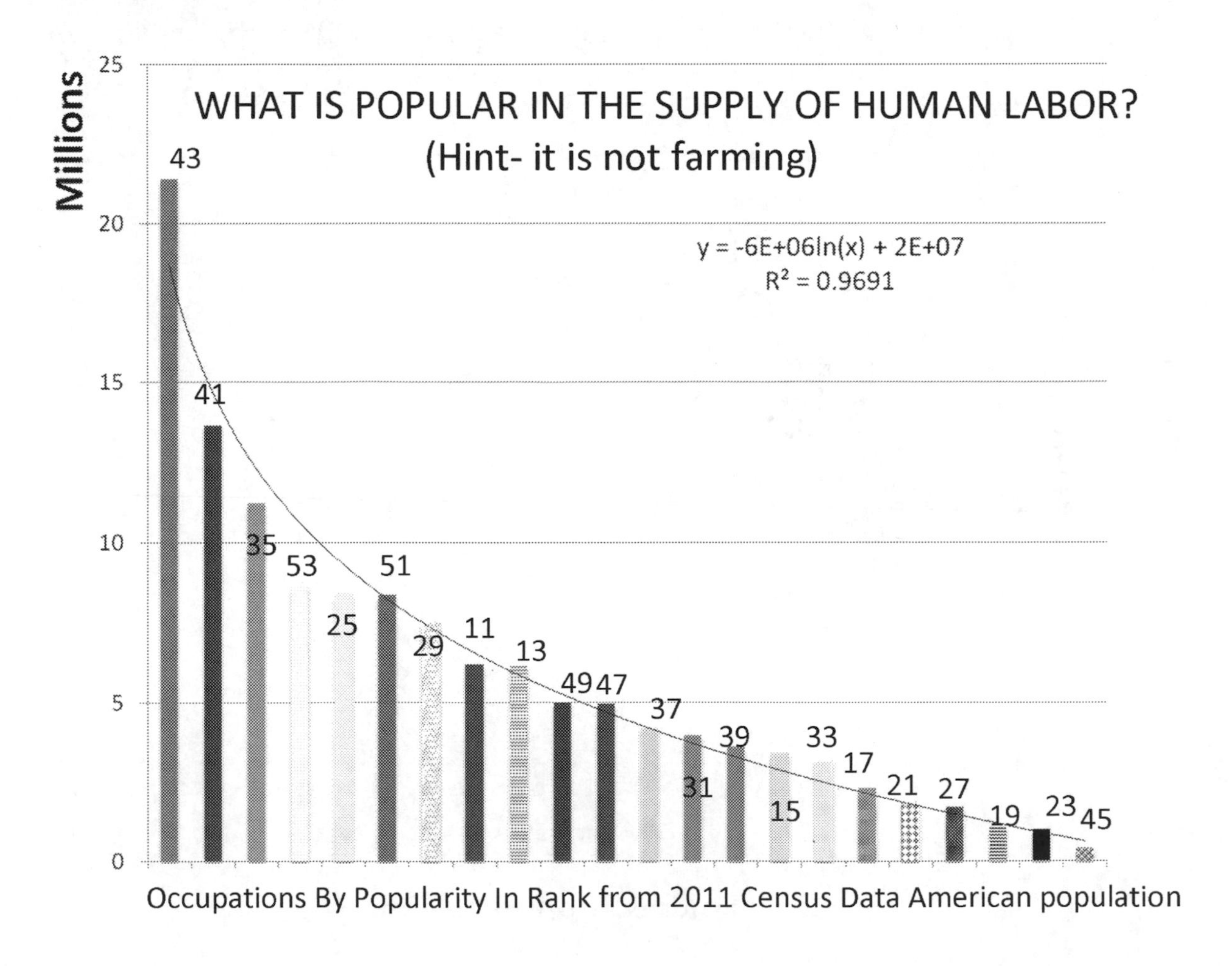

Image 4A: This shows the population distribution of occupations using the US Census Bureau occupational classification of the twenty-two categories

of industries to see what choices are popular by outcomes (the numbering scheme starts with 11 and is in increments of 2).
11=management
13=business/finance
15=computer and math
17=architecture and engineering
19=life, physical, and social sciences
21=community and social services
23=legal
25=education, training, and library
27=arts, design, entertainment, sports, and media
29=health-care practitioners and technical operations
31=health-care support
33=protective service
35=food preparation and serving
37=building, grounds, and maintenance
39=personal care and service
41=sales and related
43=office and administrative support
45=farming, fishing, and forestry
47=construction and extraction
49=installation, maintenance, and repair
51=production
53=transportation and material moving

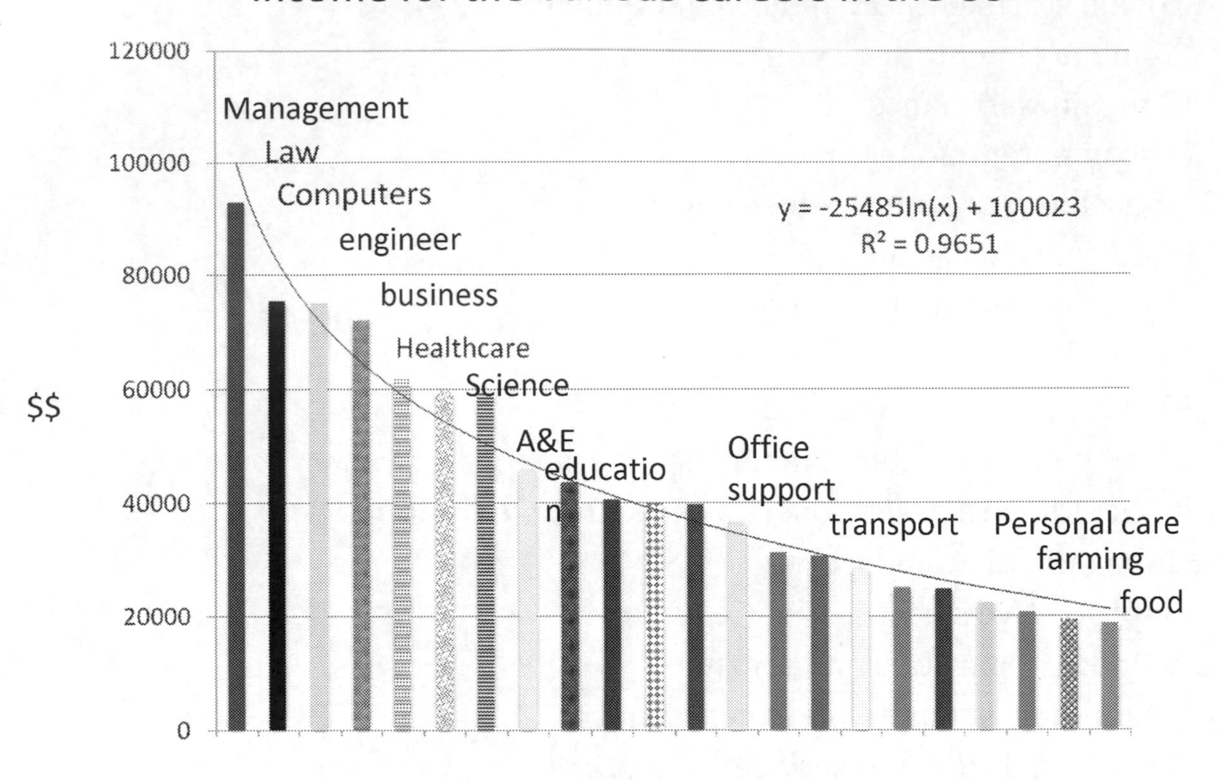

Image 5A: These are the same occupations as the previous image showing the relationship of pay (demand) with supply as shown above with the same color scheme as earlier graph. My point here is the jobs that are rewarded in today's world are thinking related (and hence education based) rather than manual labor. Data from Bureau of Labor Statistics, national_M2011_dl.xls from http://www.bls.gov/data/archived.htm.

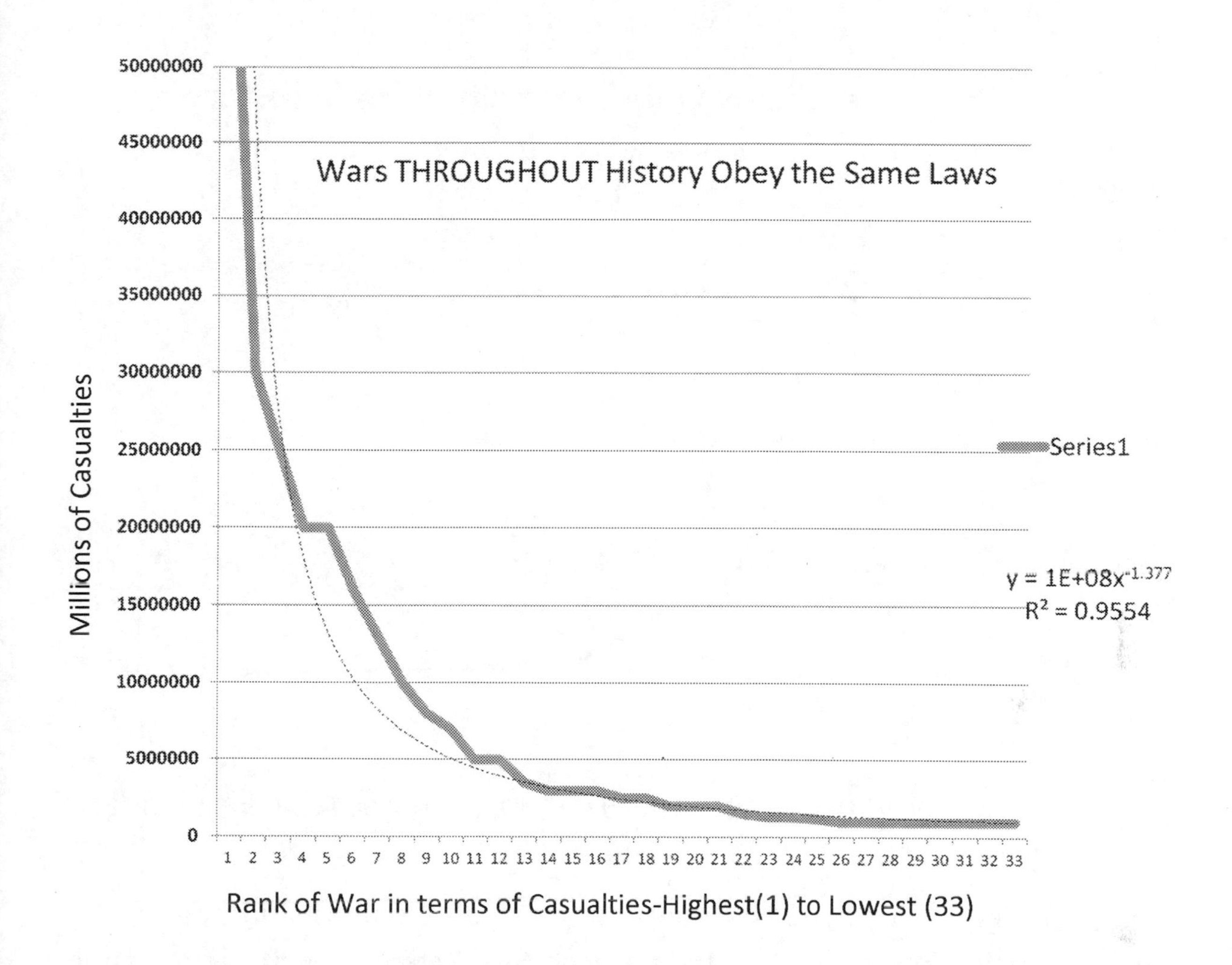

Image 6A: Wars throughout history obey the same principles of imbalance as seen elsewhere in this book in regard to human behavioral concepts. A few wars have been particularly popular at depopulating nations.

Data was graphed using estimates from *Wikipedia*, s.v. "List of Wars and Anthropogenic Disasters by Death Toll" from 2013 data entries last accessed July 2013.

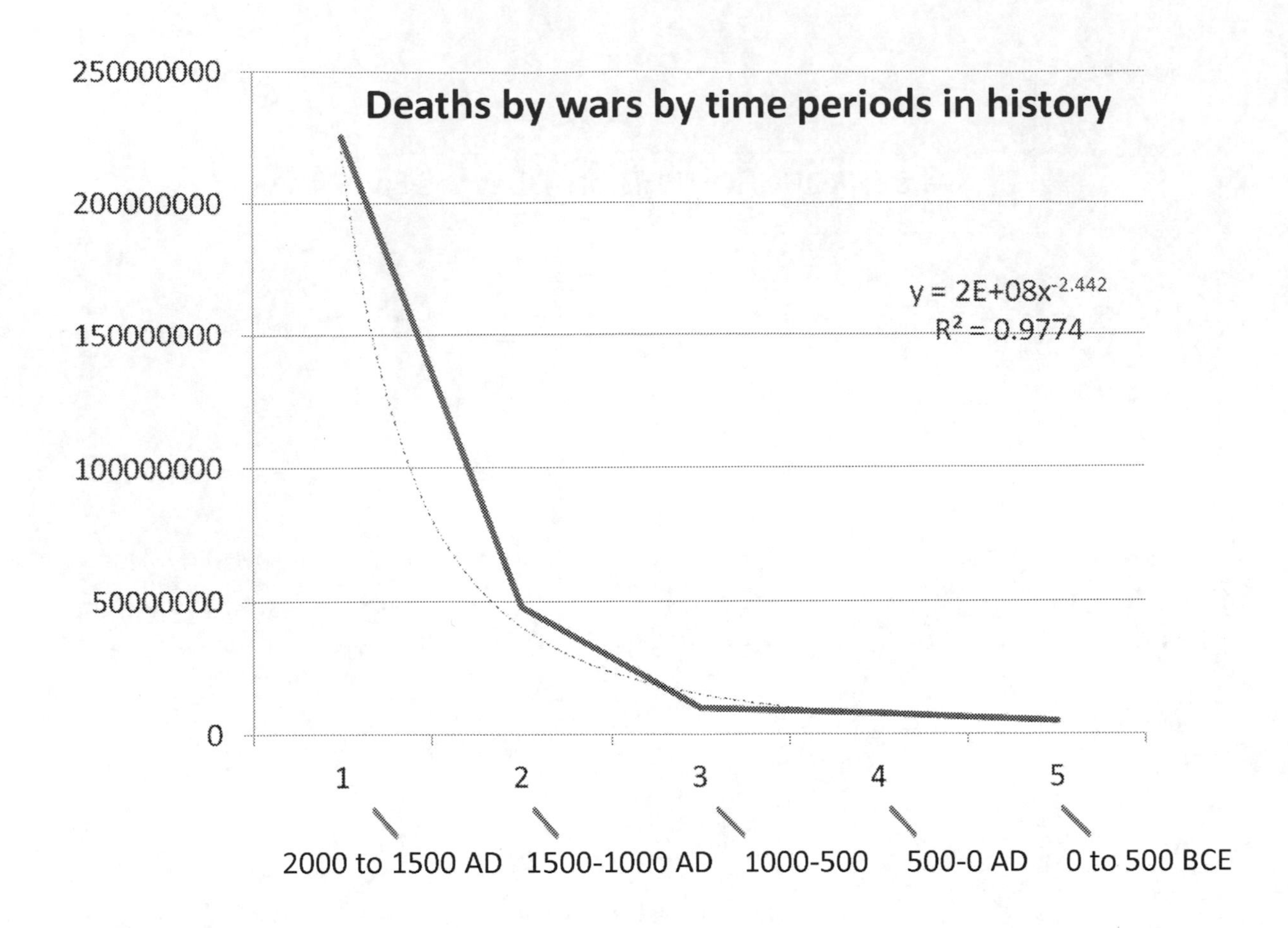

Image 7A: This shows we have become more efficient with killing in war with time. I grouped the data from image 6A into five-hundred-year time periods cumulatively to show the effect of this mostly learned efficiency that reflects what I would assume are similar inputs of technology and globalization.

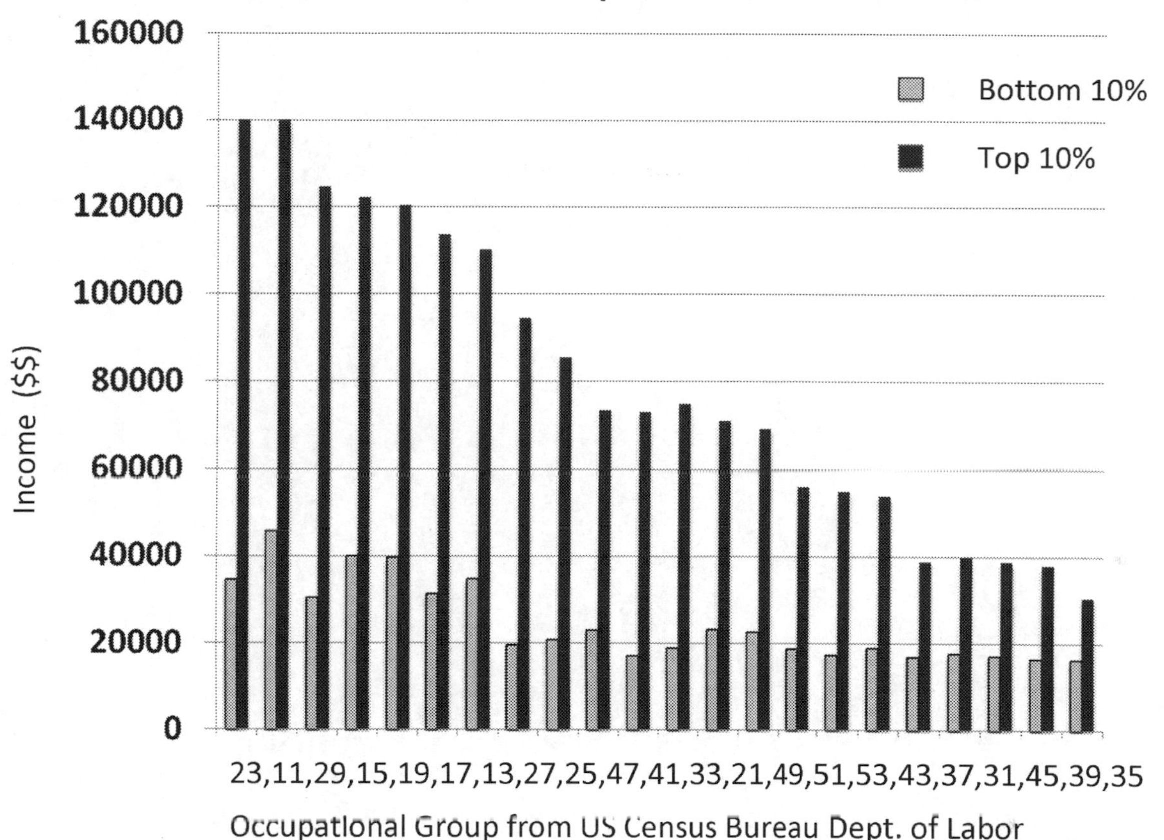

Image 8A: This image shows occupations and the separation of income to show the disparity idea a bit more relative to the various industries. The x-axis shows the number corresponding to the industry classification I used in image 4A from US Census Bureau data. (23=legal occupations, 11=management, 29=health-care practitioners, etc.).

About the Author

George LaRoque III is a physician who cares for cancer patients in rural Appalachia. He is a father of 4 and wrote this book to help teach the ideas that he sees that are important to his children's generation. He draws from his personal experiences as a physician and as a parent raising 4 very different children in the modern world.

Printed in the United States
by Baker & Taylor Publisher Services